DO DICE PLAY GOD?

Ian Stewart is Professor Emeritus of Mathematics at the University of Warwick and the author of the bestseller *Professor Stewart's Cabinet of Mathematics Curiosities*. His recent books include *Significant Figures, Incredible Numbers, Seventeen Equations that Changed the World, Professor Stewart's Casebook of Mathematical Mysteries* and *Calculating the Cosmos* (all published by Profile). His app, Incredible Numbers, was published jointly by Profile and Touch Press in 2014. He is a Fellow of the Royal Society and was the 2015 recipient of the Mathematics Association of American Euler Book Prize.

D0916487

DO DICE PLAY GOD?

The Mathematics of Uncertainty

IAN STEWART

First published in Great Britain in 2019 by
PROFILE BOOKS LTD
3 Holford Yard
Bevin Way
London WC1X 9HD
www.profilebooks.com

1 3 5 7 9 10 8 6 4 2

Printed and bound in Great Britain by
Clays Ltd, Elcograf S.p.A.

A CIP catalogue record for this book is available from the British Library.

ISBN 978 1 78125 9436
Export 978 178816 2289
eISBN 978 78283 4014

Contents

1

SIX AGES OF UNCERTAINTY

Uncertain: The state of not being definitely known or perfectly clear; doubtfulness
or vagueness.
The Oxford English Dictionary

UNCERTAINTY ISN'T ALWAYS BAD. We like surprises, as long as they're
pleasant ones. Many of us enjoy a flutter on the horses, and most
sports would be pointless if we knew at the start who was going to
win. Some prospective parents are keen *not* to be told the sex of the
baby. Most of us, I suspect, would prefer not to know in advance the
date of their own death, let alone how it will occur. But those are
exceptions. Life is a lottery. Uncertainty often breeds doubt, and doubt
makes us feel uncomfortable, so we want to reduce, or better still
eliminate, uncertainty. We worry about *what will happen*. We look out
for the weather forecast, even though we know that weather is
notoriously unpredictable and the forecast is often wrong.

When we watch the news on television, or read a newspaper, or
surf the web, the extent to which we don't know what's going to
happen can be overwhelming. Aircraft crash at random. Earthquakes
and volcanoes devastate communities, even large parts of cities. The
financial sector booms and busts, and although we speak of the 'boom
and bust cycle', all we mean is that boom follows bust and bust follows
boom. We have little idea when one of them will switch to the other.
We might as well speak of the 'rainy and dry cycle' and claim to
forecast the weather. When elections are in the offing, we keep an eye
on the opinion polls, hoping to get some inkling about who is likely to
win. Polls in recent years seem to have become less reliable, but they
still have the power to reassure or annoy us.

Sometimes we're not just uncertain; we're uncertain about what we
ought to be uncertain about. Most of us worry about climate change,

but a vocal minority insists it's all a hoax – perpetrated by scientists (who couldn't organise a hoax to save their lives), or by the Chinese, or maybe Martians ... pick your favourite conspiracy theory. But even the climatologists who predicted climate change offer few certainties about its precise effects. They do have a pretty clear handle on their general nature, though, and in practical terms that's more than enough to set alarm bells ringing.

Not only are we uncertain about what Mother Nature will throw at us; we're not too sure about what we throw at ourselves. The world's economies are still reeling from the 2008 financial crisis, while the people who caused it are mostly conducting their business as before, which is likely to bring about an even bigger financial disaster. We have very little idea how to forecast global finances.

After a period of relative (and historically unusual) stability, world politics is becoming increasingly fractured, and old certainties are being shaken. 'Fake News' is submerging genuine facts in a barrage of disinformation. Predictably, those who complain most loudly about it are often the ones responsible for the fakery. The internet, instead of democratising knowledge, has democratised ignorance and bigotry. By removing the gatekeepers, it has left the gates hanging off their hinges.

Human affairs have always been messy, but even in science, the old idea of nature obeying exact laws has given way to a more flexible view. We can find rules and models that are approximately true (in some areas 'approximate' means 'to ten decimal places', in others it means 'between ten times as small and ten times as large') but they're always provisional, to be displaced if and when fresh evidence comes along. Chaos theory tells us that even when something *does* obey rigid rules, it may still be unpredictable. Quantum theory tells us that deep down at its smallest levels, the universe is *inherently* unpredictable. Uncertainty isn't just a sign of human ignorance; it's what the world is made of.

WE COULD JUST BE FATALISTIC about the future, as many people are. But most of us are uncomfortable about that way of living. We suspect that it will probably lead to disaster, and we have a sneaking feeling that with a little foresight, disaster might be averted. A common human tactic, when faced with something we dislike, is either to guard against it, or try to change it. But what precautions should we take, when we don't know what's going to happen? After the *Titanic* disaster, ships were required to fit extra lifeboats. Their weight caused the *S.S. Eastland* to capsize on Lake Michigan, and 848 people died. The Law of Unintended Consequences can foil the best of intentions.

We're concerned about the future because we're time-binding animals. We have a strong sense of our location in time, we anticipate future events, and we act now because of those anticipations. We don't have time machines, but we often behave as if we do, so that a future event causes us to take action before it occurs. Of course the real cause of today's action isn't the wedding or the thunderstorm or the rent bill that will happen tomorrow. It's our present belief that it's going to happen. Our brains, shaped by both evolution and individual learning, let us choose our actions today to make our lives easier tomorrow. Brains are decision-making machines, making guesses about the future.

The brain makes some decisions a split second ahead. When a cricketer or baseball player catches the ball, there's a small but definite time delay between the visual system detecting the ball and the brain working out where it is. Remarkably, they usually catch the ball, because their brain is pretty good at anticipating its trajectory, but when they fumble an apparently easy catch, either the prediction or their reaction to it went wrong. The whole process is subconscious and apparently seamless, so we don't notice that we live our entire lives in a world that's running a split second ahead of our brain.

Other decisions may be taken days, weeks, months, years, or even decades ahead. We get up in time to get to the bus or train to work. We buy food for tomorrow's meals, or next week's. We plan a family outing for the coming public holiday, and everyone involved does things *now* to prepare for *then*. Wealthy parents in the UK sign their children up for the posh schools before they're born. Wealthier ones plant trees that won't mature for centuries, so that their great-great-great-grandchildren will get an impressive view.

How does the brain foretell the future? It builds simplified internal models of how the world works, or may work, or is presumed to work. It feeds what it knows into the model, and observes the outcome. If we spot a loose carpet, one of these models tells us that this could be dangerous, causing someone to trip and fall down the stairs. We take preventative action and fix the carpet in the correct position. It doesn't really matter whether this particular forecast is right. In fact, if we've fixed the carpet properly, it *can't* be right, because the conditions fed into the model no longer apply. However, evolution or personal experience can test the model, and improve it, by seeing what happens in similar cases when preventative action isn't taken.

Models of this kind need not be accurate descriptions of how the world works. Instead, they amount to *beliefs* about how the world works. And so, over tens of thousands of years, the human brain evolved into a machine that makes decisions based on its beliefs about where those decisions will lead. It's therefore no surprise that one of the earliest ways we learned to cope with uncertainty was to construct systematic beliefs about supernatural beings who were in control of nature. We knew *we* weren't in control, but nature constantly surprised us, often unpleasantly, so it seemed reasonable to assume that some inhuman entities – spirits, ghosts, gods, goddesses – *were* in control. Soon a special class of people came into being, who claimed they could intercede with the gods to help us mortals achieve our aims. People who claimed to foretell the future – prophets, seers, fortune-tellers, oracles – became especially valued members of the community.

This was the first Age of Uncertainty. We invented belief systems, which became ever more elaborate because every generation wanted to make them more impressive. We rationalised the uncertainty of nature as the will of the gods.

THIS EARLIEST STAGE OF CONSCIOUS human engagement with uncertainty lasted for thousands of years. It agreed with the evidence, because the will of the gods could credibly be whatever happened. If the gods were pleased, good things happened; if they were angry, bad things happened. As proof, if good things happened to you then you were obviously pleasing the gods, and if bad things happened it was your own fault for making them angry. So beliefs about gods became entangled with moral imperatives.

Eventually, it began to dawn on increasing numbers of people that belief systems with that flexibility didn't really *explain* anything. If the reason for the sky being blue is that the gods made it that way, it might just as well have been pink or purple. Humanity began to explore a different way of thinking about the world, based on logical inference supported (or denied) by observational evidence.

This was science. It explains blue sky in terms of scattering of light by fine dust in the upper atmosphere. It doesn't explain why blue *looks* blue; the neuroscientists are working on that, but science has never claimed to understand everything. As it grew, science achieved increasingly many successes, along with some ghastly failures, and it began to give us the ability to control some aspects of nature. The discovery of the relation between electricity and magnetism in the 19th century was one of the first truly revolutionary instances of science being turned into technology that affected the lives of almost everybody.

Science showed us that nature can be less uncertain than we think. Planets don't wander about the sky according to godly whim: they follow regular elliptical orbits, aside from tiny disturbances that they inflict on each other. We can work out which ellipse is appropriate, understand the effect of those tiny disturbances, and predict where a planet will be centuries ahead. Indeed, nowadays, millions of years, subject to the limitations imposed by chaotic dynamics. There are natural laws; we can discover them and use them to predict what will happen. The uncomfortable feeling of uncertainty gave way to the belief that most things would be explicable if we could tease out the underlying laws. Philosophers began to wonder whether the entire universe is just the working out, over aeons of time, of those laws. Maybe free will is an illusion, and it's all a huge clockwork machine.

Perhaps uncertainty is merely temporary ignorance. With enough effort and thought, all can become clear. This was the second Age of Uncertainty.

SCIENCE ALSO FORCED US TO find an effective way to quantify how certain or uncertain an event is: probability. The study of uncertainty became a new branch of mathematics, and the main thrust of this book is to examine the different ways in which we've exploited mathematics in the quest to render our world more certain. Many other things have contributed too, such as politics, ethics, and art, but I'll focus on the role of mathematics.

Probability theory grew from the needs and experiences of two very different groups of people: gamblers and astronomers. Gamblers wanted a better grasp of 'the odds', astronomers wanted to obtain accurate observations from imperfect telescopes. As the ideas of probability theory sank into human consciousness, the subject escaped its original confines, informing us not just about dice games and the orbits of asteroids, but about fundamental physical principles. Every few seconds, we breathe in oxygen and other gases. The vast number of molecules that make up the atmosphere bounce around like diminutive billiard balls. If they all piled up in one corner of the room while we were in the opposite corner, we'd be in trouble. In principle, it could happen, but the laws of probability imply that it's so rare that in practice it never does. Air stays uniformly mixed because of the second law of thermodynamics, which is often interpreted as stating that the universe is always becoming more disordered. The second law also has a somewhat paradoxical relationship to the direction in which time flows. This is deep stuff.

Thermodynamics was a relative latecomer to the scientific scene. By the time it arrived, probability theory had entered the world of human affairs. Births, deaths, divorces, suicides, crime, height, weight, politics. The applied arm of probability theory, statistics, was born. It gave us powerful tools to analyse everything from measles epidemics to how people will vote in a forthcoming election. It shed some light, though not as much as we'd like, on the murky world of finance. It told us that we're creatures afloat on a sea of probabilities.

Probability, and its applied arm statistics, dominated the third Age of Uncertainty.

THE FOURTH AGE OF UNCERTAINTY arrived with a bang, at the start of the 20th century. Until then, all forms of uncertainty that we had encountered had a common feature: uncertainty reflected human ignorance. If we were uncertain about something, it was because we didn't have the information needed to predict it. Consider tossing a coin, one of the traditional icons of randomness. However, a coin is a very simple mechanism, mechanical systems are deterministic, and in principle any deterministic process is predictable. If we knew all the forces acting on a coin, such the initial speed and direction of the toss, how fast it was spinning, and about which axis, we could use the laws of mechanics to calculate which way up it would land.

New discoveries in fundamental physics forced us to revise that view. It may be true of coins, but sometimes the information we need simply isn't available, because even nature doesn't know it. Around 1900, physicists were starting to understand the structure of matter on very small scales – not just atoms, but the subatomic particles into which atoms can be divided. Classical physics, of the kind that emerged from Isaac Newton's breakthroughs with the laws of motion and gravity, had given humanity an extensive understanding of the physical world, tested using measurements of increasingly high precision. Out of all the theories and experiments, two different ways of thinking about the world crystallised: particles and waves.

A particle is a tiny lump of matter, precisely defined and localised. A wave is like ripples on water, a disturbance that moves; more ephemeral than a particle, and extending throughout a larger region of space. Planetary orbits can be calculated by pretending the planet is a particle, because the distances between planets and stars are so gigantic that if you scale everything down to human size, planets *become* particles. Sound is a disturbance in the air that travels, even though all the air stays in pretty much the same place, so it's a wave. Particles and waves are icons of classical physics, and they're very different.

In 1678 there was a big controversy about the nature of light. Christiaan Huygens presented his theory that light is a wave to the Paris Academy of Sciences. Newton was convinced that light is a stream of particles, and his view prevailed. Eventually, after a hundred years spent barking up the wrong tree, new experiments settled the issue. Newton was wrong, and light is a wave.

Around 1900 physicists discovered the photoelectric effect: light hitting certain types of metal can cause a small electrical current to flow. Albert Einstein deduced that light is a stream of tiny particles – photons. Newton had been right all along. But Newton's theory had been discarded for a good reason: lots of experiments showed very clearly that light is a wave. The debate opened up all over again. Is light a wave, or a particle? The eventual answer was 'both'. Sometimes light behaves like a particle, sometimes like a wave. It depends on the experiment. This was all very mysterious.

A few pioneers quickly began to see a way to make sense of the puzzle, and quantum mechanics was born. All of the classical certainties, such as the position of a particle and how fast it moves, turned out not to apply to matter on subatomic scales. The quantum world is riddled with uncertainty. The more precisely you measure the position of a particle, the less sure you can be about how fast it's moving. Worse, the question 'where is it?' has no good answer. The best you can do is to describe the probability that it's located in a given place. A quantum particle isn't a particle at all, just a fuzzy cloud of probabilities.

The more deeply physicists probed the quantum world, the fuzzier everything became. They could describe it mathematically, but the mathematics was weird. Within a few decades they had become convinced that quantum phenomena are irreducibly random. The quantum world really *is* made from uncertainty, there's no missing information, and no deeper level of description exists. 'Shut up and calculate' became the watchword; don't ask awkward questions about what it all means.

WHILE PHYSICS WENT DOWN THE quantum route, mathematics blazed its own new trail. We used to think that the opposite of a random process is a deterministic one: given the present, only one future is possible. The fifth Age of Uncertainty emerged when mathematicians, and a few scientists, realised that a deterministic system can be unpredictable. This is chaos theory, the media's name for nonlinear dynamics. The development of quantum theory might have been rather different if mathematicians had made that vital discovery much earlier than they did. In fact, one example of chaos was discovered before quantum theory, but it was seen as an isolated curiosity. A coherent theory of chaos didn't appear until the 1960s and 1970s. Nevertheless, I'll tackle chaos before quantum theory, for presentational reasons.

'Prediction is very difficult, especially about the future,' said physicist Niels Bohr (or was it Yogi Berra? See, we can't even be certain of *that*).[1] It's not as funny as it sounds, because prediction is different from forecasting. Most predictions in science predict *that* an event will happen under certain conditions, but not *when*. I can predict that an earthquake happens because stresses build up in rocks, and that prediction can be tested by measuring the stresses. But that's not a method for predicting an earthquake, which requires determining, ahead of time, *when* it will happen. It's even possible to 'predict' that some event *did* happen in the past, which is a legitimate test of a theory if no one had noticed until they went back to the old records and looked. I know this is often called 'postdiction', but as far as testing a scientific hypothesis goes, it's the same thing. In 1980 Luiz and Walter Alvarez predicted that 65 million years ago an asteroid hit the Earth and killed the dinosaurs. It was a genuine prediction because, *after* making it, they could search the geological and fossil records for evidence for or against.

Observations over decades show that the sizes of beaks among some species of Darwin's finches, on the Galápagos Islands, are entirely predictable – provided you can predict the average yearly rainfall. The sizes change in lockstep with how wet or dry the years are. In dry years, seeds are harder, so bigger beaks are needed. In wet years, smaller beaks work better. Here, beak size is *conditionally* predictable. If a reliable oracle told us next year's rainfall, we could confidently predict the beak sizes. That's definitely different from the beak sizes being random. If they were, they wouldn't follow the rainfall.

It's not unusual for some features of a system to be predictable while others are unpredictable. My favourite example is astronomical. In 2004 astronomers announced that an obscure asteroid called 99942 Apophis might collide with the Earth on 13 April 2029, or if it missed in just the right way, there could be a second opportunity on 13 April 2036. One journalist (to be fair, in a humorous column) asked: How can they be so sure about the *date* when they don't know the *year*?

Stop reading and think about it. Hint: what is a year?

It's very simple. Potential collisions occur when the orbit of the asteroid intersects, or nearly intersects, that of the Earth. These orbits change slightly as time passes, affecting how closely the two bodies approach each other. If we don't have enough observations to determine the asteroid's orbit with sufficient precision, we can't be sure how close it will come to the Earth. The astronomers had enough orbital data to rule out most years over the next few decades, but not 2029 or 2036. In contrast, the date of a possible collision behaves quite differently. The Earth returns to (almost) the same location in its orbit after one year has passed. That's the definition of 'year'. In particular, our planet comes close to the intersection with the asteroid's orbit at intervals of one year; that is, on the same day every year. (Maybe one day ahead or behind if the timing is close to midnight.) As it happens, that day is 13 April for Apophis.

So Bohr or Berra was absolutely right, and his statement is really quite profound. Even when we understand how things work, in considerable detail, we may have no idea what will happen next week, next year, or next century.

WE HAVE NOW ENTERED THE sixth Age of Uncertainty, characterised by the realisation that uncertainty comes in many forms, each being comprehensible to some extent. We now possess an extensive mathematical toolkit to help us make sensible choices in a world that's still horribly uncertain. Fast, powerful computers let us analyse huge amounts of data quickly and accurately. 'Big data' is all the rage, although right now we're better at collecting it than we are at doing anything useful with it. Our mental models can be augmented with computational ones. We can perform more calculations in a second than all the mathematicians in history managed with pen and paper. By combining our mathematical understanding of the different forms that uncertainty can take, with intricate algorithms to tease out patterns and structures, or just to quantify how uncertain we are, we can to some extent tame our uncertain world.

We're much better at predicting the future than we used to be. We still get annoyed when the weather forecast tells us it's not going to rain tomorrow, and it does; but the accuracy of weather prediction has improved considerably since 1922 when the visionary scientist Lewis Fry Richardson wrote *Weather Prediction by Numerical Process*. Not only is the forecast better: it's accompanied by an assessment of the probability that it's right. When the weather website says '25% chance of rain' it means that on 25% of occasions when the same statement has been made, rain has duly fallen. If it says '80% chance of rain' then it's likely to be right four times out of five.

When the Bank of England issues forecasts of changes to the rate of inflation, it similarly provides an estimate of how reliable its mathematical modellers think the forecast is. It also found an

Percentage increase in prices on a year earlier

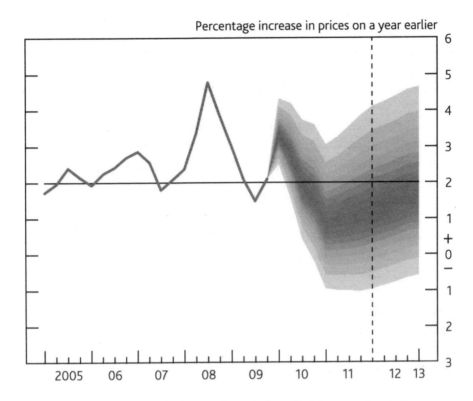

Inflation fan chart for the Bank of England's prediction of inflation according to the Consumer Price Index, February 2010.

effective way to present this estimate to the public: a 'fan chart' which plots the evolution, over time, of the predicted inflation rate, but not as a single line: as a shaded band. As time passes, the band gets wider, indicating a loss of accuracy. The density of the ink indicates the level of probability: a dark region is more likely than a fainter one. The shaded area covers 90% of the probable forecasts.

The messages here are twofold. First: as understanding advances, predictions can be made more accurate. Second: we can *manage* uncertainty by working out how confident we should be in the prediction.

A third message is also beginning to be understood. Sometimes uncertainty can actually be *useful*. Many areas of technology deliberately create controlled amounts of uncertainty, in order to make devices and processes work better. Mathematical techniques for finding the best solution to an industrial problem use random disturbances to avoid getting stuck on strategies that are the best compared to near neighbours, but not as good as more distant ones. Random changes to recorded data improve the accuracy of weather forecasts. SatNav uses streams of pseudorandom numbers to avoid problems with electrical interference. Space missions exploit chaos to save expensive fuel.

FOR ALL THAT, WE'RE STILL children 'playing on the seashore', as Newton put it, 'finding a smoother pebble or a prettier shell than ordinary, whilst the great ocean of truth lay all undiscovered before [us].' Many deep questions remain unanswered. We don't really understand the global financial system, even though everything on the planet depends on it. Our medical expertise lets us spot most disease epidemics early on, so we can take steps to mitigate their effects, but we can't always predict how they spread. Every so often new diseases appear, and we're never sure when and where the next one will strike. We can make exquisitely accurate measurements of earthquakes and volcanoes, but our track record of predicting them is as shaky as the ground beneath our feet.

The more we find out about the quantum world, the more hints there are that some deeper theory can make its apparent paradoxes more reasonable. Physicists have given mathematical proofs that quantum uncertainty can't be resolved by adding a deeper layer of reality. But proofs involve assumptions, which are open to challenge, and loopholes keep turning up. New phenomena in *classical* physics have uncanny similarities to quantum puzzles, and we know that their workings have nothing to do with irreducible randomness. If we'd known about them, or about chaos, before discovering quantum weirdness, today's theories might have been very different. Or perhaps we'd have wasted decades looking for determinism where none exists.

I've bundled everything up neatly into six Ages of Uncertainty, but the reality was less tidy. Principles that ultimately turned out to be very simple emerged in complex and confusing ways. There were unexpected twists and turns, big leaps forward, and dead ends. Some mathematical advances turned out to be red herrings; others languished for years before anyone recognised their significance. There were ideological splits, even among the mathematicians. Politics, medicine, money, and the law got in on the act, sometimes all at once.

It's not sensible to tell this kind of story in chronological order, even within individual chapters. The flow of ideas matters more than the flow of time. In particular, we'll get to the fifth Age of Uncertainty (chaos) before the fourth Age (quantum). We'll look at modern applications of statistics before we encounter older discoveries in fundamental physics. There will be diversions into curious little puzzles, a few simple calculations, and some surprises. Nevertheless, everything is here for a reason, and it all fits together.

Welcome to the six Ages of Uncertainty.

2

READING THE ENTRAILS

When the members of the household are treated with stern severity, there has been no failure. When wife and children are smirking and chattering, the economy of the family has been lost.

I Ching

WITHIN THE TOWERING WALLS of the city of Babylon, the king, resplendent in royal regalia, raises his hand. A hush falls over the crowd of nobles and officials, assembled in the huge temple courtyard.

Outside, the ordinary people go about their daily business, blissfully unaware that what's about to happen could change their lives completely. No matter: they're used to this, it's the will of the gods. There's nothing to be gained by worrying or complaining. They seldom even think about it.

The *bārû* priest waits at the sacrificial altar, knife in hand. A sheep, carefully chosen according to the ancient rituals, is led in on a short rope. The animal senses that something unpleasant is about to happen. It bleats and struggles to free itself.

The knife slashes across its throat, blood spurts. The crowd gives a collective moan. When the blood has slowed to a trickle, the priest makes a careful incision and extracts the sheep's liver. Laying it reverently on the blood-spattered stone, he bends over and studies the severed organ closely. In the crowd, people hold their breath. The king strides a few paces to the priest's side. They confer in low voices, gesturing and occasionally pointing at some feature of the excised organ – a blemish here, an unusual protrusion there. The priest places wooden pegs into holes in a special clay tablet to record their observations. Apparently satisfied, the priest confers once more with the king, then steps back respectfully while the king turns to face his nobles.

When he announces that the omens for attacking a neighbouring kingdom are favourable, they cheer in triumph. Later, on the battlefield, some may see it all rather differently, but by then it will be too late.

IT MIGHT HAVE BEEN LIKE that. We know very little about the Old Babylonian Kingdom, even at its end around 1600 BC, but something roughly along these lines must have been a common event in that ancient city. Babylon was famous for it. The Bible tells us[2] that 'the king of Babylon stands at the parting of the way, at the head of the two ways, to use divination; he shakes the arrows, he consults the household idols, he looks at the liver.' The Babylonians believed that specially trained priests, known as *bārû*, could interpret a sheep's liver to foretell the future. They compiled a huge list of omens, the *Bārûtu*. For practical reasons, and to provide quick answers, the *bārû* made do with a shorter summary in an actual divination. Their procedures were systematic and laden with tradition; they inspected specific regions of the liver, each with its own meaning, each symbolic of a particular god or goddess. The *Bārûtu* still exists as more than a hundred clay tablets incised with cuneiform writing, and it lists more than *eight thousand* omens. The wealth of information the Babylonians believed to be encoded in a single organ from a dead sheep is extraordinary in its diversity, obscurity, and occasional banality.

The *Bārûtu* has ten main chapters. The first two deal with parts of the unfortunate animal other than the liver, while the remaining eight concentrate on specific features: *be na*, the 'station', a groove on the left lobe of the liver; *be gír*, the 'path', another groove at right angles to the first; *be giš.tukul*, the 'fortuitous markings', a small protrusion, and so on. Many of these regions were further subdivided. The omens associated with each region were stated as predictions, often historical, as though the priests were recording previous associations between regions of the liver and the events that unfolded. Some were specific: 'Omen of King Amar-Su'ena, who was gored by an ox, but died from the bite of a shoe'. (This obscure statement may refer to a scorpion bite when he was wearing sandals.) Some still ring true today: 'The accountants will plunder the palace.' Others seem specific yet lack key details: 'A famous person will arrive riding on a donkey.' Others were so vague as to be practically useless: 'long-term forecast: lament'. Some regions of the liver were classified as unreliable or ambiguous. It all looks highly organised, and, in a strange way, almost scientific. The list was compiled over a long period, repeatedly edited and expanded, and copied by later scribes, which is how it has come down to us. Other evidence survives too. In particular, the British Museum has a clay model of a sheep's liver from the period 1900–1600 BC.

We now call this method of foretelling the future *hepatomancy* – liver divination. More generally, *haruspicy* is divination by the inspection of the entrails of sacrificed animals (mainly sheep and chickens), and *extiscipy* is divination using organs in general, focused on their shape and location. The methods were taken up by the Etruscans, as shown for instance by a bronze sculpture of a liver from 100 BC, found in Italy, which is divided into regions marked with the names of Etruscan deities. The Romans continued the tradition; their term for a *bārû* was *haruspex*, from *haru* = entrails, *spec-* = observe. The practice of reading the entrails is recorded in the time of Julius Caesar and Claudius, but ended under Theodosius I around 390 AD when Christianity finally displaced the last vestiges of older cults.

WHY AM I TELLING YOU this, in a book about the mathematics of uncertainty?

Divination shows that the deep human desire to foretell the future goes back a long way. Its roots are no doubt far older, but the Babylonian inscriptions are detailed, their provenance secure. History also shows how religious traditions become ever more elaborate as time passes. The records make it abundantly clear that Babylonian royalty and the priesthood believed in the method – or, at the very least, found it convenient to appear to believe. But the long interval during which hepatomancy was practised suggests that the beliefs were genuine. Even today, similar superstitions abound – avoid black cats and ladders, throw a pinch of salt over your shoulder if you spill some, a broken mirror brings bad luck. 'Gypsies' at fairgrounds still offer to read your palm and tell your fortune, for a financial consideration, and their jargon of fate lines and the girdle of Venus is reminiscent of the

Bārûtu's esoteric classification of the marks on a sheep's liver. Many of us are sceptical about such beliefs, others grudgingly concede that 'there might be something to it', and some are absolutely certain that the future can be predicted from the stars, tea leaves, lines on a human palm, tarot cards, or the casting of yarrow stalks in the Chinese *I Ching*, the Book of Changes.

Some divination techniques are elaborated in systems just as complex as the *Bārûtu* of old Babylon. *Plus ça change...* The famous person arriving on a donkey is reminiscent of the tall dark stranger of modern horoscopes in tabloid newspapers, a prediction vague enough to be associated with sufficient possible events to 'confirm' it, yet specific enough to convey an impression of arcane knowledge. Leading, of course, to a secure income for the fortune-teller.

Why are we so obsessed with predicting the future? It's sensible and natural, because we've always lived in an uncertain world. We still do, but at least we now have some understanding of why and how our world is uncertain, and to some extent we can put this knowledge to good use. Our forebears' world was less certain. An earthquake was not anticipated as slippage of rock along a geological fault, which can be monitored for a dangerous level of stress. It was a chance act of nature, whose unpredictability was attributed to the whims of powerful supernatural beings. At the time, this was the simplest and perhaps the only way to make sense of events that happened at random, with no obvious reason. *Something* must cause them, and it had to be something with a will of its own, able to decide they should happen, and with the power to ensure that they did. A god or a goddess was the most plausible explanation. Deities had power over nature; they did whatever they wanted to, when they wanted to, and ordinary mortals were stuck with the consequences. At least, with gods, there was some prospect of propitiating them and influencing what they did – or so the priesthood maintained, and there was no advantage in questioning authority, let alone disobeying. Anyway, the right magical rituals, the prerogative of royalty and the priesthood, might open a window into the future and resolve some of the uncertainty.

Behind all this was an aspect of the human condition that arguably singles out our species from most other animals: time-binding. We're conscious that there will be a future, and we plan our current behaviour in the context of our expectations of that future. Even when we were hunter-gathers on the African savannahs, the tribal elders knew that the seasons would turn, animals would migrate, and different plants would be ready for use at different times. Distant signs in the sky heralded a coming storm, and the earlier you noticed them, the more chance you had of taking shelter before the storm arrived. By anticipating the future, you could sometimes mitigate some of its worst effects.

As societies and their technologies became more advanced, we bound time ever more actively, with increasing accuracy and reach. Nowadays we get up at a specific time during the working week, *because* we want to catch the local train, to get to our workplace. We know the time that the train should leave from the station; we know when it's supposed to arrive at its destination; we arrange our lives to get us to work on time. Anticipating the arrival of the weekend, we book tickets to the football, the movies, or the theatre. We reserve a table at a restaurant several weeks ahead because it's going to be Esmerelda's birthday on Saturday the 29th. We buy Christmas cards in the January sales because they're cheaper then, and put them away for the next eleven months until we need them. Then we desperately try to remember where we put them. In short, our lives are heavily influenced by events that we think are going to happen in the future. It would be difficult to explain our actions without taking that into account.

As time-binding creatures, we know that the future doesn't always work out as expected. The train to work is late. A thunderstorm knocks out our internet connection. A hurricane sweeps through and devastates a dozen Caribbean islands. An election doesn't turn out the way the polls forecast, and our lives are turned upside down by people with whom we profoundly disagree. Not surprisingly, we place considerable value on predicting the future. It helps us protect ourselves and our families, and it gives us a feeling (however illusory) of being in control of our destiny. So desperate are we to know what the future holds that we're suckers for one of the oldest con-tricks in the book – people who claim to have special knowledge of future events. If a priest can influence a god, he can arrange for a favourable future. If a shaman can predict when the rains will come, at least we can get ready in advance without wasting too much time waiting. If a clairvoyant can cast our horoscope, we can keep an eye open for that tall dark stranger or donkey-borne celebrity. And if any of them can convince us that their abilities are genuine, we will flock to use their services.

Even if it's a load of old tosh.

WHY DO SO MANY OF us still believe in luck, destiny, omens?

Why are we so easily impressed by arcane symbols, long lists, complicated words, elaborate and archaic costumes, rituals, and chants?

Why do we fondly imagine that the vast, incomprehensible universe gives a toss about a bunch of overdeveloped apes on a damp lump of rock circling a very ordinary star, just one of the *ten to the power seventeen* (a hundred quadrillion) stars in the observable region of a presumably even vaster universe? Why do we interpret that universe in human terms? Is it even the kind of entity that *can* give a toss?

Why do we so readily believe obvious nonsense, even today?

Of course I'm talking about *your* beliefs, not mine. Mine are rational, firmly grounded in factual evidence, the outcome of ancient wisdom that guides me to live my life the way everyone should. Yours are mindless superstition, lack any factual basis, are supported only by unquestioning deference to tradition, and you keep telling everybody else how they should behave.

You think much the same about me, of course, but there's a difference.

I'm right.

That's the trouble with belief. Belief in the sense of blind faith is often inherently untestable. Even when it is testable, we often ignore the results, or, if the test disproves our beliefs, we deny its significance. This attitude may be irrational, but it reflects the evolution and organisation of the human brain. From inside any given human mind, beliefs make sense. Even ones that the outside world considers silly. Many neuroscientists think that the human brain can sensibly be thought of as a Bayesian decision machine (Thomas Bayes was a Presbyterian minister and an adept statistician – more about him in Chapter 8). Roughly, I'm referring to a device whose very structure is the embodiment of beliefs. Through personal experiences and longer-term evolution, our brains have assembled a network of interlocking assumptions about how likely some event is, given some other event. If you hit your thumb with a hammer, it will hurt: likelihood pretty much certain. If you go out in the rain without a raincoat or umbrella, you'll get wet: ditto. If the sky looks grey but it's currently dry, and you go out without a raincoat or umbrella, you'll get wet: well, probably not. Aliens visit the Earth regularly in UFOs (unidentified flying objects – flying saucers): certain, if you're a believer; definitely not, if you aren't.

When we run into fresh information, we don't just accept it. We'd be crazy to: the evolution of the human brain has been heavily influenced by the need to distinguish fact from fiction, truth from lies. We judge new information in the context of what we already believe. Someone claims to have seen a strange light in the sky, moving impossibly fast? Clear evidence of an alien visitation, if you believe in UFOs. Misinterpretation, or possibly plain invention, if you don't. We make such judgements instinctively, often without reference to the actual evidence.

Some of us might grapple with contradictions, as the rational aspect of our brain notices apparent inconsistencies. Some tormented souls lose their faith altogether. Others become converts to a new religion, cult, belief system ... call it what you will. But most of us stick pretty closely to what we were brought up to believe. The 'epidemiology' of religion, the way membership of specific sects changes over generations, shows that you catch your beliefs from parents, siblings, relatives, teachers, and authority figures in your culture. And that's one reason why we often hold strong beliefs that outsiders consider to be rubbish. If you're brought up to worship the cat goddess, warned every day of the dire consequences that will befall you if you forget to burn the holy incense or chant the proper spells, these acts and the accompanying feeling of satisfaction soon become ingrained. In fact, they're being wired into your Bayesian decision-making brain, and it may become impossible for you to disbelieve, no matter how contradictory other evidence might appear. Just as a bell push wired to a doorbell can't suddenly decide to start the car instead. That would take drastic rewiring, and rewiring a brain is extremely difficult. Moreover, knowing which spells to chant distinguishes your culture from all those barbarians who don't even believe in the cat goddess, let alone worship her.

Beliefs are also easy to reinforce. Positive evidence can always be found if you keep looking and are selective. So many things happen every day, some good, some bad: among them will be events that strengthen your beliefs. Your Bayesian brain tells you to ignore the rest: they don't matter. It filters them out. Which is why there's so much fuss about Fake News. The problem is, it does matter. But it takes an extra dose of supercharging for your rational mind to overrule those wired-in assumptions.

I was once told that on Corfu there's a superstition that when you see a praying mantis, it either brings good luck or bad luck, *depending on what happens*. This might sound ludicrous (and it might not be true), but when survivors of today's natural disasters thank God for hearing their prayers and saving their lives, it seldom occurs to them that those who died are no longer present to complain. Some Christian sects interpret a praying mantis as a symbol of piety; others, as a symbol of death. I suppose it depends on why you think the mantis is praying, and of course on your beliefs (or not) about prayer.

Humans evolved to function effectively in a chaotic world. Our brains are crammed with quick-and-dirty solutions to potential problems. Does breaking a mirror really bring bad luck? Experimenting by smashing every mirror in sight is expensive; it achieves little if the superstition is wrong, but we're asking for trouble if it's right. So much simpler to avoid breaking one, just in case. Every decision of this kind strengthens a link in the network of probabilities of the Bayesian brain.

In the past, these links served us well. It was a simpler world, and we lived a simpler lifestyle. If we occasionally fled in panic from a leopard that turned out to be a bush swaying in the breeze, at worst we looked a bit silly. But today, if too many of us try to run the planet on our beliefs, without respecting objective evidence, we'll inflict serious damage on ourselves and everyone else.

IN HIS TEENS, THE PSYCHOLOGIST Ray Hyman started reading palms to earn money. He didn't believe in it, to begin with, but he had to pretend he did, or the clients wouldn't materialise. He followed traditional interpretations of the lines on palms, and after a while his predictions became so successful – as reported by his clients – that he started to believe there must be something to it after all. Stanley Jaks, a professional mentalist who knew all the tricks of the trade, suggested that as an experiment Hyman should work out what the lines on his clients' palms meant, and then tell them *the exact opposite*. He did, and 'to my surprise and horror, my readings were just as successful as ever'. Hyman promptly became a sceptic.[3]

His clients didn't, of course. They subconsciously selected the predictions that seemed accurate, and ignored the ones that were wrong. With everything being vague and ambiguous anyway, hence open to interpretation, believers could find plenty of evidence that palmistry worked. The Corfu superstition about a praying mantis *always* works, because no subsequent event can possibly disprove it.

Precisely why some ancient civilisations attached so much importance to a sheep's liver is somewhat mysterious, but hepatomancy is just one weapon in the futurologist's extensive armoury. As Ezekiel, recorded, the king of Babylon also consulted the household idols – asked the gods. And, a literal weapon, he 'shook the arrows'. This is *belomancy*. After the Babylonian era, it found favour with Arabs, Greeks, and Scythians. There were several ways to do it, but all of them used special ritual arrows decorated with magical symbols. Arcane symbolism is always impressive, especially to the less educated: it hints at secret powers, hidden knowledge. Possible answers to a question of high import were written down and tied to different arrows, which were then shot into the air. Whichever answer travelled the furthest was the correct one. Or, perhaps to avoid wasting time tracking down distant arrows, they were simply placed in a quiver and one was drawn at random.

Livers, arrows – what else? Just about anything. Gerina Dunwich's *The Concise Lexicon of the Occult* lists a hundred different methods of divination. We're familiar with *horoscopy*, predicting a person's fate from the configuration of stars at their birth; with *cheiromancy*, better known as *palmistry*, reading their future from the lines on their palm; and with *tasseography* – tea leaves. But these merely scratch the surface of humanity's ability to imagine how the unfolding of the universe might be predicted from everyday objects. If you're not into reading palms, why not have a go at *podomancy*, predicting a person's destiny from the lines on their feet? Or *nephelomancy*: inferring future events from the shapes and directions of clouds. *Myomancy*: divination from the squeaks of mice or rats. *Sycomancy* – do it with figs. *Cromniomancy*: onion sprouts. Or you can go the whole hog (rather goat or ass) with *kephalonomancy*, once much used by Germans and Lombards. Sacrifice said goat or ass, remove the head, and bake it. Pour lighted carbon on to the head while reading out the names of suspected criminals.[4] When the head makes a crackling sound, you've identified the guilty party. Not forecasting the future, this time, but digging out a secret from the past.

At first glance, these methods are so disparate that it's hard to see any common ground, other than performing some kind of rite with everyday ingredients and decoding the arcane meaning of whatever happens. However, many of these methods depend on the same presumption: to understand something large and complicated, mimic it with something *small* and complicated. The shapes that tea leaves make in a cup are varied, random, and unpredictable. The future is *also* varied, random, and unpredictable. It's not a huge leap of logic to suspect that the two might be linked. Ditto clouds, mouse utterances, and lines on your feet. If you believe in destiny, your fate is predetermined at your birth – so why shouldn't it be written somewhere for an adept to read? What changes with the date and time of your birth? The motion of the Moon and planets across the backdrop of fixed stars... *Aha!*

It's not just ancient cultures, which lacked the extensive scientific knowledge we now have. Many people still believe in astrology. Others don't exactly *believe*, but they find it fun to read their horoscopes and see if they turn out to be right. People in many countries, in vast numbers, play the national lottery. They know the chance of winning is very small (they may not appreciate just *how* small), but you've got to be in it to win it, and if you do, your financial worries disappear in an instant. I'm not arguing that it's sensible to play, because almost everybody loses, but I know someone who won half a million pounds...

The lottery (which takes similar forms in many nations) is a game of pure chance, a view supported by statistical analysis, but thousands of players imagine that some clever system can beat the odds.[5] You can buy a miniature lottery machine that spits out tiny numbered balls at random. Use that to choose what to bet on. Inasmuch as there's a rationale, it must be along the lines of 'the lottery machine works just like the miniature one, both are random, therefore in some mysterious way the miniature one behaves like the real one'. The large is repeated in the small. It's the same logic as tea leaves and squeaky mice.

23

3

ROLL OF THE DICE

The best throw of the dice is to throw them away.
Sixteenth-century proverb

FOR MILLENNIA, HUMANITY'S WISH TO predict the future manifested itself as innumerable methods for divination, oracular pronouncements, elaborate ceremonies to propitiate the gods, and a great deal of superstition. Very little of this had any impact on rational thinking, let alone scientific or mathematical content. Even if anyone thought of making records of predictions and comparing them to events, there were too many ways to brush aside inconvenient data – the gods would be offended, you misinterpreted the oracle's advice. People routinely fell into the trap of confirmation bias: noticing anything that agrees with the prediction or belief, and ignoring anything that doesn't. They still do, in their hundreds of millions, today.

However, there was one area of human activity where ignoring facts automatically led to disaster, and that was gambling. Even there, there's room for a certain amount of self-deception; millions of people still have irrational and incorrect beliefs about probability. But millions more have a pretty good grasp of odds and how they combine, so gamblers and bookmakers make more profit when they understand the basics of probability. Not necessarily as formal mathematics, but as a well-honed grasp of the basics plus a few rules of thumb and deductions from experience. Unlike religious predictions (oracles and so on) and political ones (confident assertions with no factual basis, Fake News, propaganda), gambling provides an objective test of your beliefs about probability: whether, in the long run, you make money or lose it. If your widely touted system for beating the odds doesn't work,

24

you quickly find out and regret trying it. You can sell it to credulous mugs, but that's different. If *you* use it, with your own money, reality quickly bites.

Gambling is, and has long been, big business. Every year, worldwide, about *ten trillion* dollars changes hands in legal gambling. (A lot more if you include the financial sector...) Much of it is to some extent recycled: the punter bets on the horses, the bookie pays out on winnings but keeps the stakes from losing bets; much of the money passes through many people's hands, flowing this way and that; but in the long run a significant fraction ends up in the pockets (and bank accounts) of the bookies and casinos, and stays there. So the net amount of cash that ends up as profit, though substantial, is somewhat smaller.

The first genuine mathematics of probability theory appeared when mathematicians started to think carefully about gambling and games of chance, especially the likely behaviour in the long term. The pioneers of probability theory had to extract sensible mathematical principles from the confused mass of intuitions, superstitions, and quick-and-dirty guesswork that had previously been the human race's methods for dealing with chance events. It's seldom a good idea to start by tackling social or scientific problems in all their complexity. If the early mathematicians had tried to forecast the weather, for instance, they wouldn't have got very far, because the methods available were inadequate. Instead, they did what mathematicians always do: they thought about the simplest examples, where most of the complexity is eliminated and it's possible to specify, clearly, what you're talking about. These 'toy models' are often misunderstood by everyone else, because they seem far removed from the complexities of the real world. But throughout history, major discoveries vital to the progress of science have emerged from toy models.[6]

THE ARCHETYPAL ICON OF CHANCE is a classic gambling device: dice.[7]

Dice may have originated in the Indus valley, arising from the even older use of knucklebones – animal bones used for fortune-telling and playing games. Archaeologists have found six-sided dice, essentially the same as those in use today, at Shahr-e Sūkhté (Burnt City) in ancient Iran, a site that was occupied from 3200 to 1800 BC. The oldest dice date from about 2800–2500 BC, and were used to play a game similar to backgammon. At much the same time, the ancient Egyptians used dice to play the game of *senet*, whose rules are unknown, though there's no shortage of guesses.

We can't be sure whether these early dice were used for gambling. The ancient Egyptians didn't have money as such, but they often used grain as a form of currency, part of an elaborate barter system. But gambling with dice was common in Rome, two millennia ago. There's something strange about most Roman dice. At first sight they look like cubes, but nine tenths of them have rectangular faces, not square ones. They lack the symmetry of a genuine cube, so some numbers would have turned up more frequently than others. Even a slight bias of this kind can have a big effect in a long series of bets, which is how dice games are normally played. Only in the middle of the 15th century did it become standard to use symmetric cubes. So why didn't Roman gamblers object when they were asked to play with biased dice? Jelmer Eerkens, a Dutch archaeologist who has made a study of dice, wondered whether a belief in fate, rather than physics, might be the explanation. If you thought your destiny was in the hands of the gods, then you'd win when they wanted you to win and lose when they didn't. The shape of the dice would be irrelevant.[8]

By 1450, gamblers seem to have wised up, because most dice were symmetric cubes by then. Even the arrangement of the numbers had become standardised, possibly to make it easier to check that all six numbers appeared. (A standard way to cheat, still in use today, is to secretly swap the dice for rigged ones on which some numbers occur twice, making them more likely to be thrown. Put them on opposite faces and a cursory glance won't spot the trick. Do this with two dice and some totals are impossible. There are dozens of other ways to cheat, even with completely normal dice.) Initially, most dice had 1 opposite 2, 3 opposite 4, and 5 opposite 6. This arrangement is known as primes, because the totals 3, 7, and 11 are all prime. Around 1600, primes went out of favour and the configuration we still use today took over: 1 opposite 6, 2 opposite 5, and 3 opposite 4. This is called 'sevens' since opposite faces always add to 7. Both primes and sevens can occur in two distinct forms, which are mirror images of each other.

It's possible that as dice became more regular and standardised, gamblers adopted a more rational approach. They stopped believing that Lady Luck could influence a biased dice, and paid more attention to the likelihood of any particular outcome occurring without divine intervention. They could hardly have failed to notice that with an unbiased dice, although the numbers thrown don't appear in any predictable sequence, any given number is just as likely as the others. So in the long run each should turn up equally often, give or take some variation. This kind of thinking eventually led some pioneering mathematicians to create a new branch of their subject: probability theory.

THE FIRST OF THESE PIONEERS was Girolamo Cardano, in Renaissance Italy. He won his mathematical spurs in 1545 by writing *Ars Magna*, the Great Art. This was the third really important book on what we now call algebra. The *Arithmetica* of the Greek mathematician Diophantus, dating from around 250 AD, introduced symbols to represent unknown numbers. The *al-Kitāb al-mukhtaṣar fī ḥisāb al-jabr wal-muqābala* (The Compendious Book on Calculation by Completion and Balancing) by the Persian mathematician Mohammed al-Khwarizmi around 800 AD gave us the word 'algebra'. He didn't use symbols, but he developed systematic methods for solving equations – 'algorithms', from the Latin version of his name, Algorismus. Cardano put both ideas together – symbolic representation of the unknown, plus the possibility of treating the symbols as new kinds of mathematical objects. He also went beyond his predecessors by solving more elaborate equations.

His mathematical credentials were impeccable, but his character left much to be desired: he was a gambler and a rogue, with a tendency to violence. But he lived in an age of gamblers and rogues, where violence was often just around the next corner. Cardano was also a doctor, quite a successful one by the standards of the period. And he was an astrologer. He got into trouble with the Church by casting Christ's horoscope. It's said that he cast his own horoscope, which got him into even more trouble because, having predicted the date of his own death, his professional pride led him to kill himself to make sure the prediction came true. There seems to be no objective evidence for this story, but given Cardano's personality, it has a certain inevitability.

Before looking at Cardano's contribution to probability theory, it's worth sorting out some terminology. If you bet on a horse, the bookmaker doesn't offer you a probability: he quotes you *odds*. For instance, he may offer odds of 3:2 on Galloping Girolamo in the 4.30 Renaissance Stakes at Fakingham Racecourse. What that means is that if you bet £2 and win, the bookie gives you back £3, *plus* your original stake of £2. If you win, you're £3 better off; if you lose, the bookie is £2 better off.

This arrangement is fair, in the long run, if the wins and losses cancel each other out. So a horse at odds of 3:2 ought to lose three times for every two wins. In other words, there should, on average, be two wins in every five races. The *probability* of a win is therefore two out of five: 2/5. In general, if the odds are *m*:*n*, the probability of the horse winning is

$$p = \frac{n}{m+n}$$

if the odds are scrupulously fair. They seldom are, of course; bookies are in business to make a profit. On the other hand, the odds will be close to this formula: bookies don't want customers to realise they're being ripped off.

Whatever the practicalities, this formula tells you how to convert odds into probabilities. You can also go the other way, bearing in mind that odds are a ratio: 6:4 is the same as 3:2. The ratio *m*:*n* is the fraction *m*/*n*, which is equal to $1/p - 1$. As a check, if $p = 2/5$ we get $m/n = 5/2 - 1 = 3/2$, odds of 3:2.

CARDANO WAS ALWAYS SHORT OF ready cash, and augmented his finances by becoming an expert gambler and chess player. His *Liber de ludo aleae* (Book on Games of Chance) was written in 1564, but not published until 1663 as part of his collected works, by which time he was long dead. It includes the first systematic treatment of probability. He used dice to illustrate some basic concepts, and wrote: 'To the extent to which you depart from ... equity, if it is in your opponent's favour, you are a fool, and if in your own, you are unjust.' This is his definition of 'fair'. Elsewhere in the book he explains how to cheat, so it looks like he didn't actually *object* to injustice as long as it happened to someone else. On the other hand, even an honest gambler needs to know how to cheat, to spot opponents doing it. On the basis of this remark he explains why fair odds can be viewed as the ratio of losses to wins (for the punter, or wins to losses for the bookie). In effect, he defines the probability of an event to be the proportion of occasions on which it happens, in the long run. And he illustrated the mathematics by applying it to gambling with dice.

He prefaced his analysis with the statement that 'the most fundamental principle of all in gambling is simply equal conditions, e.g. of opponents, of bystanders, of money, of situation, of the dice box, and of the dice itself'. With this convention, the roll of a single dice is straightforward. There are six outcomes, and if the dice is fair, each occurs on average once every six throws. So the probability of each is 1/6. When it comes to two or more dice, Cardano got the basics right when several other mathematicians got them wrong. He stated that there are 36 equally likely throws of two dice and 216 throws of three. Today we would observe that $36 = 6 \times 6$ (or 6^2) and $216 = 6 \times 6 \times 6$ (or 6^3), but Cardano did the sums like this: 'There are six throws with like faces, and fifteen combinations with unlike faces, which when doubled gives thirty, so that there are thirty-six throws in all.'

Why double? Suppose one dice is red and the other blue. Then the combination of a 4 and a 5 can occur in two different ways: red 4 and blue 5; red 5 and blue 4. However, the combination of a 4 and a 4 occurs in just one way: red 4 and blue 4. The colours are introduced here to make the reasoning transparent: even if the dice *look* identical, there are still two ways to roll a combination of different numbers, but only one way to roll the same number with both dice. The crucial ingredients are ordered pairs of numbers, not unordered pairs.[9] Simple though this observation may seem, it was a significant advance.

For three dice, Cardano solved a long-standing conundrum. Gamblers had long known from experience that when throwing three dice, a total of 10 is more likely than 9. This puzzled them, however, because there are six ways to get a total of 10:

$$1+4+5 \quad 1+3+6 \quad 2+4+4 \quad 2+2+6 \quad 2+3+5$$

$$3+3+4$$

but *also* six ways to get a total of 9:

$$1+2+6 \quad 1+3+5 \quad 1+4+4 \quad 2+2+5 \quad 2+3+4$$
$$3+3+3$$

So why does 10 occur more often? Cardano pointed out that there are 27 *ordered* triples totalling 10, but only 25 totalling 9.[10]

He also discussed throwing dice many times repeatedly, and there he made his most significant discoveries. The first was that the probability of the event is the proportion of occasions on which it happens, in the long run. This is now known as the 'frequentist' definition of probability. The second was that the probability of an event occurring every time in n trials is p^n if the probability of a single event is p. It took him a while to get the right formula, and his book included the mistakes he made along the way.

YOU WOULDN'T EXPECT A LAWYER and a Catholic theologian to have much interest in gambling, but Pierre de Fermat and Blaise Pascal were accomplished mathematicians who couldn't resist a challenge. In 1654 the Chevalier de Méré, renowned for his gambling expertise, which apparently extended 'even to the mathematics' – a rare compliment indeed – asked Fermat and Pascal for a solution to the 'problem of the points'.

Consider a simple game in which each player has a 50% chance of winning, for example, tossing a coin. At the start, they contribute equal stakes to a 'pot', and agree that the first person to win some specific number of rounds ('points') wins the money. However, the game is interrupted before it finishes. Given the scores at that stage, how should the gamblers divide the stakes? For example, suppose the pot is 100 francs, the game is supposed to stop as soon as one player wins 10 rounds, but they have to abandon it when the score is 7 to 4. How much should each player get?

This question triggered an extensive correspondence between the two mathematicians, which still survives except for Pascal's initial letter to Fermat, where apparently he suggested the wrong answer.[11] Fermat responded with a different calculation, urging Pascal to reply and say whether he agreed with the theory. The answer was as he'd hoped:

> Monsieur,
>
> Impatience has seized me as well as it has you, and although I am still abed, I cannot refrain from telling you that I received your letter in regard to the problem of the points yesterday evening from the hands of M. Carcavi, and that I admire it more than I can tell you. I do not have the leisure to write at length, but, in a word, you have found the two divisions of the points and of the dice with perfect justice.

Pascal admitted his previous attempt had been wrong, and the two of them batted the problem to and fro, with Pierre de Carcavi (like Fermat, a mathematician and a parliamentary counsellor) acting as intermediary. Their key insight is that what matters is not the past history of the play – aside from setting up the numbers – but what might happen over the remaining rounds. If the agreed target is 20 wins and the game is interrupted with the score 17 to 14, the money ought to be divided in exactly the same way as it would be for a target of 10 and scores 7 to 4. (In both cases, one player needs 3 more points and the other needs 6. How they reached that stage is irrelevant.) The two mathematicians analysed this set-up, calculating what we would now call each player's expectation – the average amount they would win if the game were to be repeated many times. The answer for this example is that the stakes should be divided in the ratio 219 to 37, with the player in the lead getting the larger part. Not something you'd guess.[12]

THE NEXT IMPORTANT CONTRIBUTION CAME from Christiaan Huygens in 1657, with *De ratiociniis in ludo aleae* (On Reasoning in Games of Chance). Huygens also discussed the problem of points, and made explicit the notion of expectation. Rather than writing down his formula, let's do a typical example. Suppose you play, many times, a dice game where your wins or losses are:

lose £4 if you throw 1 or 2
lose £3 if you throw 3
win £2 if you throw 4 or 5
win £6 if you throw 6

It's not immediately clear whether you have an advantage in the long run. To find out, calculate:

the probability of losing £4 is $2/6 = 1/3$
the probability of losing £3 is $1/6$
the probability of winning £2 is $2/6 = 1/3$
the probability of winning £6 is $1/6$

Then, says Huygens, your expectation is obtained by multiplying each win or loss (losses counting as negative numbers) by the corresponding probability, and adding them up:

$$\left(-4 \times \frac{1}{3}\right) + \left(-3 \times \frac{1}{6}\right) + \left(2 \times \frac{1}{3}\right) + \left(6 \times \frac{1}{6}\right)$$

which is $-1/6$. That is, on average you lose $16\frac{2}{3}$ pence per game.

To see why this works, imagine that you toss the dice six million times, and each number comes up a million times – the average case. Then you may as well toss the dice six times, with each number comes up once, because the proportions are the same. In those six tosses you lose £4 when you throw 1 and 2, you lose £3 when you throw 3, you win £2 when you throw 4 and 5, and you win £6 when you throw 6. Your total 'winnings' are therefore

$$(-4) + (-4) + (-3) + 2 + 2 + 6 = -1$$

If you divide by 6 (the number of games concerned) and group together terms with the same loss or win, you reconstruct Huygens's expression. The expectation is a sort of average of the individual wins or losses, but each outcome must be 'weighted' according to its probability.

Huygens also applied his mathematics to real problems. With his brother Lodewijk, he used probabilities to analyse life expectancy, based on the tables of John Graunt published in 1662 in *Natural and Political Observations Made upon the Bills of Mortality*, generally held to be the earliest significant work on demography, the study of population numbers, and one of the earliest on epidemiology, the study of epidemic diseases. Already, probability and human affairs were becoming intertwined.

4

TOSS OF A COIN

Heads I win, tails you lose.
Often said in children's games

ALL OF THE PREVIOUS WORK on probabilities pales into insignificance compared to Jakob Bernoulli's epic *Ars conjectandi* (The Art of Conjecture), which he wrote between 1684 and 1689. It was published posthumously in 1713 by his nephew Nicolaus Bernoulli. Jakob, who had already published extensively on probability, collected together the main ideas and results then known, and added many more of his own. The book is generally considered to mark the arrival of probability theory as a branch of mathematics in its own right. It begins with the combinatorial properties of permutations and combinations, which we'll revisit shortly in modern notation. Next, he reworked Huygens's ideas on expectation.

Coin tossing is staple fodder for probability texts. It's familiar, simple, and illustrates many fundamental ideas well. Heads/tails is the most basic alternative in games of chance. Bernoulli analysed what's now called a *Bernoulli trial*. This models repeatedly playing a game with two outcomes, such as a coin toss with either H or T. The coin can be biased: H might have probability 2/3 so that T has probability 1/3, for instance. The two probabilities must add up to 1, because each toss is either a head or a tail. He asked questions like 'What's the probability of getting at least 20 Hs in 30 tosses?', and answered them using counting formulas known as permutations and combinations. Having established the relevance of these combinatorial ideas, he then developed their mathematics in considerable depth. He related them to the binomial theorem, an algebraic result expanding powers of the 'binomial' (two-term expression) $x + y$; for example,

$$(x + y)^4 = x^4 + 4x^3y + 6x^2y^2 + 4xy^3 + y^4$$

The third portion of the book applies the previous results to games of cards and dice that were common at that period. The fourth and last continues the emphasis on applications, but now to decision-making in social contexts, including law and finance. Bernoulli's big contribution here is the law of large numbers, which states that over a large number of trials, the number of times that any specific event H or T occurs is usually very close to the number of trials multiplied by the probability of that event. Bernoulli called this his golden theorem, 'a problem in which I've engaged myself for twenty years'. This result can be viewed as a justification of the frequentist definition of probability: 'proportion of times a given event happens'. Bernoulli saw it differently: it provided a theoretical justification for using proportions in experiments to deduce the underlying probabilities. This is close to the modern axiomatic view of probability theory.

Bernoulli set the standard for all who followed, but he left several important questions open. One was practical. Calculations using Bernoulli trials became very complicated when the number of trials is large. For example, what is the probability of getting 600 or more Hs in 1000 tosses of a fair coin? The formula involves multiplying 600 whole numbers together and dividing by another 600. In the absence of electronic computers, performing the sums by hand was at best tedious and time-consuming, and at worst beyond human abilities. Solving questions like this constituted the next big step in the human understanding of uncertainty via probability theory.

DESCRIBING THE MATHEMATICS OF PROBABILITY theory in historical terms becomes confusing after a while, because the notation, terms, and even concepts changed repeatedly as mathematicians groped their way towards a better understanding. So at this point I want to explain some of the main ideas that came out of the historical development in more modern terms. This will clarify and organise some concepts we need for the rest of the book.

It seems intuitively clear that a fair coin will, in the long run, produce about the same number of heads as tails. Each individual toss is unpredictable, but cumulative results over a series of tosses are predictable on average. So although we can't be certain about the outcome of any particular toss, we can place limits on the amount of uncertainty in the long run.

I tossed a coin ten times, and got this sequence of heads and tails:

T H T T T H T H H T

There are 4 heads and 6 tails – nearly an equal split, but not quite. How likely are those proportions?

I'll work up to the answer gradually. The first toss is either H or T, with equal probabilities of 1/2. The first two tosses could have been any of HH, HT, TH, or TT. There are four possibilities, all equally likely, so each has probability 1/4. The first three tosses could have been any of HHH, HHT, HTH, HTT, THH, THT, TTH, or TTT. There are eight possibilities, all equally likely, so each has probability 1/8. Finally, let's look at the first four tosses. There are 16 sequences, each with probability 1/16, and I'll list them according to how many times H occurs:

0 times	1 sequence (TTTT)
1 time	4 sequences (HTTT, THTT, TTHT, TTTH)
2 times	6 sequences (HHTT, HTHT, HTTH, THHT, THTH, TTHH)
3 times	4 sequences (HHHT, HHTH, HTHH, THHH)
4 times	1 sequence (HHHH)

The sequence I tossed began THTT, with only one head. That number of heads occurs 4 times out of 16 possibilities: probability $4/16 = 1/4$. For comparison, two heads and two tails occurs 6 times out of 16 possibilities: probability $4/16 = 3/8$. So although heads and tails are equally likely, the probability of getting the same number of each isn't 1/2; it's smaller. On the other hand, the probability of getting *close* to two heads – here 1, 2, or 3 – is $(4 + 6 + 4)/16 = 14/16$, which is 87·5%.

With ten tosses there are $2^{10} = 1024$ sequences of Hs and Ts. Similar sums (there are short cuts) show that the numbers of sequences with a given number of Hs go like this:

0 times	1 sequence	probability 0·001
1 time	10 sequences	probability 0·01
2 times	45 sequences	probability 0·04
3 times	120 sequences	probability 0·12
4 times	210 sequences	probability 0·21
5 times	252 sequences	probability 0·25
6 times	210 sequences	probability 0·21
7 times	120 sequences	probability 0·12
8 times	45 sequences	probability 0·04
9 times	10 sequences	probability 0·01
10 times	1 sequence	probability 0·001

My sequence has 4 heads and 6 tails, an event with probability 0·21. The most probable number of heads is 5, with probability only 0·25. Picking a particular number of heads isn't terribly informative. A more interesting question is: What is the probability of getting a number of heads and tails in some range, such as somewhere between 4 and 6? The answer is 0·21 + 0·25 + 0·21 = 0·66. In other words, if we toss a coin ten times, we can expect either a 5:5 or 6:4 split two times out of three. The flipside is that we can expect a *greater* disparity one time out of three. So a certain amount of fluctuation around the theoretical average is not only possible, but quite likely.

If we look for a larger fluctuation, say a split of 5:5, 6:4, or 7:3 (either way), the probability of staying within those limits becomes 0·12 + 0·21 + 0·25 + 0·21 + 0·12 = 0·9. Now the chance of getting a worse imbalance is about 0·1 – one in ten. This is small, but not ridiculously so. It's surprising that when you toss a coin ten times, the chance of getting two or fewer heads, or two or fewer tails, is as big as 1/10. It will happen, on average, once every ten trials.

AS THESE EXAMPLES ILLUSTRATE, EARLY work in probability mainly focused on counting methods for equally likely cases. The branch of mathematics that counts things is known as combinatorics, and the concepts that dominated the earliest work were permutations and combinations.

A permutation is a way to arrange a number of symbols or objects in order. For example, the symbols A B C can be ordered in six ways:

ABC   ACB   BAC   BCA   CAB   CBA

Similar lists show that there are 24 ways to arrange four symbols, 120 ways to arrange five, 720 ways to arrange six, and so on. The general rule is simple. For instance, suppose we want to arrange six letters A B C D E F in some order. We can select the first letter in six different ways: it's either A or B or C or D or E or F. Having chosen the first letter, we're left with five others to continue the ordering process. So there are five ways to choose the second letter. Each of those can be appended to the initial choice, so overall we can choose the first two letters in

$$6 \times 5 = 30$$

ways. There are four choices for the next letter, three choices for the next letter, two choices for the next letter, and the sixth letter is the only one left. So the total number of arrangements is

$$6 \times 5 \times 4 \times 3 \times 2 \times 1 = 720$$

The standard symbol for this calculation is 6!, read as 'six factorial' (or more properly as 'factorial six', but hardly anyone ever says that).

By the same reasoning, the number of ways to arrange a pack of 52 cards in order is

$$52! = 52 \times 51 \times 50 \times \cdots \times 3 \times 2 \times 1$$

which my faithful computer tells me, with impressive speed, is

80,658,175,170,943,878,571,660,636,856,403,766,975,289,505,440, 883,277,824,000,000,000,000

This answer is exact, gigantic, and not something you could find by listing all the possibilities.

More generally, we can count how many ways there are to arrange in order any four letters from the six letters A B C D E F. These arrangements are called permutations (of four letters from six). The sums are similar, but we stop after choosing four letters. So we get

$$6 \times 5 \times 4 \times 3 = 360$$

ways to arrange four letters. The neatest way to say this

mathematically is to think of it as

$$(6 \times 5 \times 4 \times 3 \times 2 \times 1)/(2 \times 1) = 6!/2! = 720/2 = 360.$$

Here we divide by 2! to get rid of the unwanted $\times 2 \times 1$ at the end of 6!. By the same reasoning, the number of ways to arrange 13 cards out of 52 in order is

$$52!/39! = 3,954,242,643,911,239,680,000$$

Combinations are very similar, but now we count not the number of arrangements, but the number of different choices, ignoring order. For instance, how many different hands of 13 cards are there? The trick is to count the number of permutations first, and then ask how many of them are the same except for the ordering. We've already seen that every set of 13 cards can be ordered in 13! ways. That means that each (unordered) set of 13 cards appears 13! times in the (hypothetical) list of all 3,954,242,643,911,239,680,000 ordered lists of 13 cards. So the number of unordered lists is

$$3,954,242,643,911,239,680,000/13! = 635,013,559,600$$

and that's the number of different hands.

In a probability calculation, we might want the probability of getting a particular set of 13 cards – say, all the spades. That's exactly one set out of those 635 billion hands, so the probability of being dealt that hand is

$$1/635,013,559,600 = 0 \cdot 000000000001574 \ldots$$

which is 1·5 trillionths. Across the planet, it should happen once in about 635 billion hands, on average.

There's an informative way to write this answer. The number of ways to choose 13 cards out of 52 (the number of *combinations* of 13 out of 52) is

$$\frac{52!}{13!39!} = \frac{52!}{13!(52-13)!}$$

Algebraically, the number of ways to choose r objects from a set of n

objects is

$$\frac{n!}{r!(n-r)!}$$

So we can calculate this number in terms of factorials. It is often spoken informally as '*n* choose *r*'; the fancy term is *binomial coefficient*, written symbolically as

$$\binom{n}{r}$$

This name arises because of a connection with the binomial theorem in algebra. Look at my formula for $(x+y)^4$ a few pages back. The coefficients are 1, 4, 6, 4, 1. The same numbers occur when we count how many ways a given numbers of heads arises in four consecutive tossings. The same goes if you replace 4 by any whole number.

THUS ARMED, LET'S TAKE ANOTHER look at the list of 1024 sequences of Hs and Ts. I said that there are 210 sequences in which H appears four times. We can calculate this number using combinations, though how to do it isn't immediately obvious since it's related to ordered sequences in which the symbols can repeat, which looks like a very different animal. The trick is to ask at which *positions* the four Hs appear. Well, they might be in positions 1, 2, 3, 4 – HHHH followed by six Ts. Or they might be in positions 1, 2, 3, 5 – HHHTH followed by five Ts. Or... Whatever the positions may be, the list of those positions at which the four Hs appear is a list of four numbers from the full set 1, 2, 3, ..., 10. That is, it's the number of combinations of 4 numbers out of 10. But we know how to find that: we just work out

$$\frac{10!}{4!(10-4)!} = \frac{10!}{4!6!} = 210$$

number of sequences

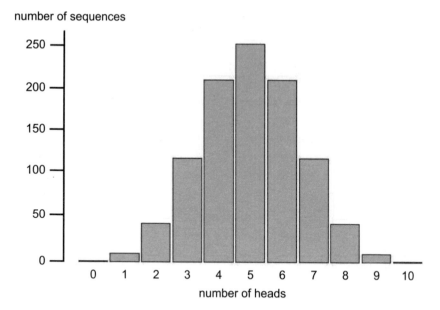

Binomial distribution for ten trials, H and T equally likely. Divide the numbers on the vertical scale by 1024 to get probabilities.

Magic! Repeating this kind of calculation, we get the entire list:

$$\frac{10!}{0!10!} = 1 \qquad \frac{10!}{1!9!} = 10 \qquad \frac{10!}{2!8!} = 45 \qquad \frac{10!}{3!7!} = 120 \qquad \frac{10!}{4!6!} = 210$$

$$\frac{10!}{5!5!} = 252$$

after which the numbers repeat in reverse order. You can see this symbolically, or argue that (for instance) six Hs is the same as four Ts, and the number of ways to get four Ts is obviously the same as the number of ways to get four Hs.

The general 'shape' of the numbers is that they start small, rise to a peak in the middle, and then decrease again, with the entire list being symmetric about the middle. When we graph the number of sequences against the number of heads, as a bar chart, or histogram if you want to be posh, we see this pattern very clearly.

A measurement made at random from some range of possible events is called a random variable. The mathematical rule that associates each value of the random variable to its probability is called a probability distribution. Here the random variable is 'number of heads', and the probability distribution looks much like the bar chart, except that the numbers labelling the vertical scale must be divided by 1024 to represent probabilities. This particular probability distribution is called a binomial distribution because of the link to binomial coefficients.

Distributions with different shapes occur when we ask different questions. For example, with one dice, the the throw is either 1, 2, 3, 4, 5 or 6, and each is equally likely. This is called a uniform distribution.

If we throw two dice and add the resulting numbers together, the totals from 2 to 12 occur in different numbers of ways:

$2 = 1 = 1$	1 way
$3 = 1 + 2, 1 + 1$	2 ways
$4 = 1 + 3, 2 + 2, 3 + 1$	3 ways
$5 = 1 + 4, 2 + 3, 3 + 2, 4 + 1$	4 way
$6 = 1 + 5, 2 + 4, 3 + 3, 4 + 2, 5 + 1$	5 ways
$7 = 1 + 6, 2 + 5, 3 + 4, 4 + 3, 5 + 2, 6 + 1$	6 ways

increasing by 1 at each step, but then the numbers of ways start to decrease because throws of 1, 2, 3, 4 and 5 are successively eliminated:

$8 = 2 + 6, 3 + 5, 4 + 4, 5 + 3, 6 + 2$	5 ways
$9 = 3 + 6, 4 + 5, 5 + 4, 6 + 3$	4 ways
$10 = 4 + 6, 5 + 5, 6 + 4$	3 ways
$11 = 5 + 6, 6 + 5$	2 ways
$12 = 6 + 6$	1 way

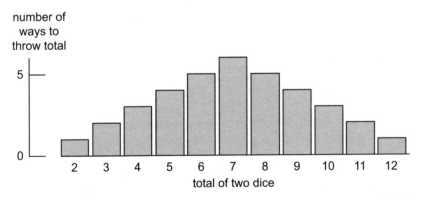

Distribution for the total of two dice. Divide the vertical scale by 36 to get probabilities.

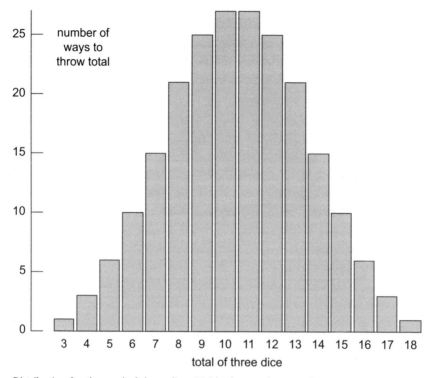

Distribution for the total of three dice. Divide the numbers on the vertical scale by 216 to get probabilities.

The probability distribution for these totals is therefore shaped like a triangle. The numbers are graphed in the figure; the corresponding probabilities are these numbers divided by their total, which is 36.

If we throw three dice and add up the resulting numbers, the shape rounds off and starts to look more like a binomial distribution, although it's not quite the same. It turns out that the more dice we throw, the closer the total gets to a binomial distribution. The central limit theorem in Chapter 5 explains why this happens.

COINS AND DICE ARE COMMON metaphors for randomness. Einstein's remark that God does not play dice with the universe is widely known. It's less widely known that he didn't use those exact words, but what he did say made the same point: he didn't think the laws of nature involve randomness. It's therefore sobering to discover that he may have chosen the wrong metaphor. Coins and dice have a dirty secret. They're not as random as we imagine.

In 2007 Persi Diaconis, Susan Holmes, and Richard Montgomery investigated the dynamics of coin tossing.[13] They started with the physics, building a coin-tossing machine that flips the coin into the air, so that it rotates freely until it lands on a flat receptive surface, without bouncing. They set up the machine to perform the flip in a very controlled way. So controlled that, as long as you place the coin heads up in the machine, it always lands heads – despite turning over many times in mid-air. Place it tails up, and it always lands tails. This experiment demonstrates very clearly that coin tossing is a predetermined mechanical process, not a random one.

Joseph Keller, an applied mathematician, had previously analysed a special case: a coin that spins around a perfectly horizontal axis, turning over and over until it's caught in a human hand. His mathematical model showed that provided the coin spins fast enough, and stays in the air long enough, only a small amount of variability in the initial conditions leads to equal proportions of heads and tails. That is, the probability that the coin comes down heads is very close to the expected value 1/2, and so is that of tails. Moreover, those figures still apply even if you always start it heads up, or tails up. So a really vigorous flip does a good job of randomisation, provided the coin spins in the special way Keller's model assumes.

At the other extreme, we can imagine a flip that is just as vigorous, but makes the coin spin around a vertical axis, like a turntable for playing vinyl records. The coin goes up, comes down again, but never flips right over, so it always lands exactly the same way up as it was when it left the hand. A realistic coin toss lies somewhere in between, with a spin axis that's neither horizontal nor vertical. If you don't cheat, it's probably close to horizontal.

Suppose for definiteness we always start the toss with heads on top. Diaconis's team showed that unless the coin obeys Keller's assumptions *exactly*, and flips about a precisely horizontal axis (which in practice is impossible) it will land with heads on top more than half the time. In experiments with people tossing it in the usual manner, it lands heads about 51% of the time and tails 49% of the time.

Before we get too worried about 'fair' coins not being fair, we must take three other factors into account. Humans can't toss the coin with the same precision as a machine. More significantly, people don't toss coins by always starting heads up. They start with heads or tails, at random. This evens out the probabilities for ending up heads or tails, so the outcome is (very close to) fifty-fifty. It's not the *toss* that creates the equal probabilities; it's the unconscious randomisation performed by the person tossing the coin when they place it on their thumb before flipping it. If you wanted to gain a slight edge, you could practise precision tossing until you got really good, and always start the coin the same way up that you want it to land. The usual procedure neatly avoids this possibility by introducing yet another element of randomness: one person tosses, and the other one calls 'heads' or 'tails' while the coin is in the air. Since the person tossing doesn't know in advance what the other person will call, they can't affect the chances by choosing which way up to start the coin.

THE ROLL OF A DICE is more complicated, with more possible outcomes. But it seems reasonable to examine the same issue. When you roll a dice, what's the most important factor in determining which face ends up on top?

There are plenty of possibilities. How fast it spins in the air? How many times it bounces? In 2012 Marcin Kapitaniak and colleagues developed a detailed mathematical model of a rolling dice, including factors like air resistance and friction.[14] They modelled the dice as a perfect mathematical cube with sharp corners. To test the model, they made high-speed movies of rolling dice. It turned out that none of the above factors is as important as something far simpler: the initial position of the dice. If you hold the dice with face 1 on top, it rolls 1 slightly more often than anything else. By symmetry, the same goes for any other number.

The traditional 'fair dice' assumption is that each face has probability $1/6 = 0.167$ of being thrown. The theoretical model shows that in an extreme case where the table is soft and the coin doesn't bounce, the face on top at the start ends up on top with probability 0.558 – much larger. With the more realistic assumption of four or five bounces, this becomes 0.199 – still significantly bigger. Only when the dice rotates very fast, or bounces about twenty times, does the predicted probability become close to 0·167. Experiments using a special mechanical device to toss the dice with a very precise speed, direction, and initial position showed similar behaviour.

5

TOO MUCH INFORMATION

A reasonable probability is the only certainty.
Edgar Watson Howe, *Sinner Sermons*

CARDANO'S *LIBER DE LUDO ALEAE* peeked inside Pandora's box. Bernoulli's *Ars conjectandi* blew the lid off. Probability theory was a game-changer – quite literally, given its uses in gambling – but its radical implications for assessing the likelihood of chance events took a long time to sink in. Statistics, which can roughly be described as the applied branch of probability theory, arrived on the scene much more recently. Some important 'prehistory' occurred around 1750. The first major breakthrough happened in 1805.

Statistics originated in two very different areas: astronomy and sociology. Their common feature was the problem of extracting useful information from imperfect or incomplete observational data. Astronomers wanted to find the orbits of planets, comets, and similar bodies. The results allowed them to test mathematical explanations of celestial phenomena, but there were also potential practical consequences, notably for navigation at sea. The social applications came a little later, with the work of Adolphe Quetelet in the late 1820s.

There was a link: Quetelet was an astronomer and meteorologist at the Royal Observatory in Brussels. But he also acted as a regional correspondent for the Belgian bureau of statistics, and it was this that made his scientific reputation. Quetelet deserves a chapter to himself, and I'll describe his ideas in Chapter 7. Here I'll concentrate on the astronomical origins of statistics, which laid the foundations of the subject so firmly that some of the methods that emerged remain in use to this day.

IN THE 18TH AND 19TH CENTURIES, the main focus of astronomy was on the motion of the Moon and planets, later extended to comets and asteroids. Thanks to Newton's theory of gravity, astronomers could write down very accurate mathematical models of many kinds of orbital motion. The main scientific problem was to compare those models with observations. Their data were obtained using telescopes, with increasing precision as the decades passed and instruments became more refined. But it was impossible to measure the positions of stars and planets with complete accuracy, so all observations were subject to uncontrollable errors. Changes in temperature affected the instruments. Refraction of light by the Earth's ever-changing atmosphere made the images of planets wobble. The screw threads used to move various scales and gauges were slightly imperfect, and when you twiddled the knobs to move them, they could stick for a moment before responding. If you repeated the same observation with the same instrument, you would often get a slightly different result.

As the engineering of the instruments improved, the same problems continued, because astronomers were always pushing the boundaries of knowledge. Better theories required observations to be ever more precise and ever more accurate. One feature ought to have operated in the astronomers' favour: their ability to make many observations of the same celestial body. Unfortunately, the mathematical techniques of the time weren't adapted to this; it seemed that *too much* data caused more problems than it solved. Actually, what the mathematicians knew, although true, turned out to be misleading; their methods solved the wrong problem. They rose to the occasion by finding new methods, as did the astronomers, but it took a while for the new ideas to sink in.

The two main techniques that misled the mathematicians were the solution of algebraic equations and the analysis of errors, both well established at that time. At school we all learn how to solve 'simultaneous equations', such as

$$2x - y = 3 \qquad 3x + y = 7$$

with the answer $x = 2$, $y = 1$. Two equations are needed to pin down the values of x and y, because one equation just relates them to each other. With three unknowns, three equations are needed to get a unique answer. The pattern extends to more unknowns: we need the same number of equations as unknowns. (There are also some technical conditions to rule out pairs of equations that contradict each other, and I'm talking about 'linear' equations where we don't see things like x^2 or xy, but let's not go into those.)

The worst feature of the algebra is that when you have *more* equations than unknowns, there's usually no solution. The jargon is that the unknowns are 'overdetermined' – you've been told too much about them, and what you've been told is contradictory. For example, if in the above calculations we also require $x + y = 4$, we're in trouble, because the other two equations already imply that $x = 2$ and $y = 1$, so $x + y = 3$. Oops. The only way an extra equation can fail to cause a problem is if it's a consequence of the first two. We'd be fine if the third equation actually were $x + y = 3$, or something equivalent like $2x + 2y = 6$, but not if the sum were anything else. Of course, this is unlikely unless the extra equation was chosen to be like that in the first place.

Error analysis focuses on a single formula, say $3x + y$. If we know that $x = 2$ and $y = 1$, then the expression equals 7. But suppose all we know is that x lies between 1·5 and 2·5, while y lies between 0·5 and 1·5. What can we say about $3x + y$? Well, in this case the largest value arises when we use the largest possibilities for x and y, leading to

$$3 \times 2{\cdot}5 + 1{\cdot}5 = 9$$

Similarly the smallest value arises when we use the smallest possibilities for x and y, leading to

$$3 \times 1{\cdot}5 + 0{\cdot}5 = 5$$

So we know that $3x + y$ lies in the range 7 ± 2. (Here $\pm$ means 'plus or minus', and the range of possible values goes from $7 - 2$ to $7 + 2$.) In fact, we can get this result more simply by just combining the largest and the smallest *errors*:

$$3 \times 0{\cdot}5 + 0{\cdot}5 = 2 \qquad 3 \times (-0{\cdot}5) - 0{\cdot}5 = -2$$

The mathematicians of the 18th century knew all this, along with more complicated formulas for errors when numbers are multiplied or divided, and how negative numbers affect the estimates. The formulas were derived using calculus, the most powerful mathematical theory of the period. The message they got from all this work is that when you combine several numbers that are subject to errors, the errors in the result get *worse*. Here, errors of ± 0.5 in x and y alone lead to errors of ± 2 in $3x + y$, for instance.

IMAGINE YOURSELF, THEN, TO BE a leading mathematician of the period, faced with 75 equations in 8 unknowns. What would you immediately 'know' about this problem?

You'd know you were in deep trouble. There are 67 more equations than you need to find the eight unknowns. You could quickly check whether you could just solve eight of the equations, and the answers would (miraculously) fit the other 67. Very accurate observations would have been consistent with each other (if the theoretical formula were correct), but these numbers were observations, and there would inevitably be errors. Indeed, in the case I have in mind, the answers to the first eight equations didn't agree with the other 67. They may have come close, but that wasn't good enough. In any case, there are nearly 17 billion ways to choose 8 equations out of 75: which should you choose?

Combining the equations to reduce their number would be one possible tactic – but the conventional wisdom was that combining equations *increases* errors.

This scenario actually happened. The mathematician was Leonhard Euler, one of the all-time greats. In 1748 the French Academy of Sciences announced its annual prize problem in mathematics. Two years earlier the astronomer Edmond Halley, famous for the eponymous comet, had noticed that Jupiter and Saturn cause each other alternately to speed up a little in their orbits, or slow down, compared to what you would expect if the other body were absent. The prize topic was to use the law of gravity to explain this effect. Euler competed, as he often did, and submitted his results in a 123-page memoir. Its main theoretical result was an equation connecting eight 'orbital elements': quantities associated with the orbits of the two bodies. To compare his theory with observations, he had to find the values of those orbital elements. There was no lack of observations: he found 75 in the astronomical archives, obtained between 1652 and 1745.

So: 75 equations in 8 unknowns – massively overdetermined. What did Euler do?

He started by manipulating the equations to obtain values for two of the unknowns, about which he was fairly confident. He achieved this by noticing that the data tended to repeat fairly closely every 59 years. So the equations for 1673 and 1732 (which are 59 years apart) looked very similar, and when he subtracted one from the other, only two significant unknowns remained. The same thing happened with data from 1585 and 1703 (which are 118 years apart: twice 59), with the same two unknowns remaining. Two equations, two unknowns: no problem there. Solving the two equations, he deduced those two unknowns.

That left him with the same 75 equations, now in only 6 unknowns – but if anything that made the problem worse: the equations were even more overdetermined. He tried the same trick on the rest of the data, but he couldn't find any combinations where most of the unknowns disappeared. Despondently he wrote: 'From these equations we can conclude nothing; and the reason, perhaps, is that I have tried to satisfy several observations exactly, whereas I should have only satisfied them approximately; and *this error has then multiplied itself* [my italics].' At this point Euler was clearly referring to the well-known fact from error analysis: combining equations magnifies errors.

After that, he messed around rather ineffectually, and achieved very little. Statistician and historian Stephen Stigler[15] remarks: 'Euler ... was reduced to groping for solutions,' and contrasts Euler's groping with the analysis performed by the astronomer Johann Tobias Mayer in 1750. Although we usually say that the Moon always presents the same face to the Earth, this is a slight simplification. Most of the far side always remains hidden, but various phenomena in effect make the bit we see wobble slightly. These wobbles are called *libration*, and that's what interested Mayer.

For roughly a year in 1748–49, Mayer observed the positions of a number of features of the lunar landscape, notably the crater Manilius. In a paper of 1750 he deduced several features of the Moon's orbit by writing down a formula with three unknowns and using his data to calculate them. He faced the same problem that had stumped Euler, because he had observations made on 27 days: 27 equations in 3 unknowns. He tackled it very differently. He split his data into three groups of nine observations, and added all the equations for each group together to get a combined equation for that group. Then he had three equations in three unknowns: not overdetermined, so he just solved them in the usual manner.

This procedure seems a bit arbitrary: how do you select the three groups? Mayer had a systematic way to do that, based on grouping together equations that looked fairly similar. It was pragmatic, but sensible. It avoided one big problem in this kind of work, numerical instability. If you solve a lot of equations that are very similar to each other, you end up dividing big numbers by small ones, causing potential errors to become quite large. He was aware of this, because he says: 'The advantage [of his choice of groups] consists in the fact that ... the differences between these three sums are made as large as possible. The greater these differences are, the more accurately one may determine the unknown values.' Mayer's method seems so sensible that we may not appreciate how revolutionary it was. No one had done anything like it before.

Hang on though: surely combining nine equations together *magnifies* the errors? If the errors are much the same for each equation, doesn't this procedure multiply the overall error by 9? Mayer certainly didn't think so. He argued that 'values ... derived from nine times as many observations ... are nine times more correct'. That is, the likely error is *divided* by 9, not multiplied.

Did he make a mistake? Or was classical error analysis wrong?

The answer is: a bit of both. The statistical point here is that error analysis, as practised at that time, focuses on the *worst-case* results, when all the individual errors combine together to give the biggest possible total. But this solves (correctly) the wrong problem. What astronomers needed was the *typical* or *most likely* overall error, and that usually involves errors of opposite signs, which to some extent cancel each other out. For example, if you have ten observations, each taking the value 5 ± 1, each observation is either 4 or 6. Their sum ranges between 40 and 60, an error of 10 compared to the correct sum of 50. In practice, though, about half of the observations will be 4 and the other half 6. If it's exactly half of each, the sum is 50 again – spot on. If it's, say, six 4s and four 6s, the sum is 48, which isn't bad. It's wrong by only 4%, when all the individual errors are wrong by 20%.

Mayer had the right idea, but he was wrong about one technical detail: his claim that nine times as many observations divides the error by 9. Later statisticians discovered that we should divide by 3 – the *square root* of 9 (see a little later). But he was on the right track.

MAYER'S METHOD FOR DEALING WITH overdetermined equations was more systematic than Euler's (which doesn't really count as a method at all), and it included the key insight that combining observations in the right way can improve accuracy, not worsen it. We owe the definitive technique to Adrien Marie Legendre, who in 1805 published A short book *Nouvelles méthodes pour la détermination des orbites des comètes* (New Methods for Determining Orbits of Comets). Legendre rephrased the problem: given an overdetermined system of linear equations, which values of the unknowns satisfies those equations *with the least overall error*?

This approach changes the game completely, because you can always find a solution if you let the error get large enough. Errors can't be eliminated completely, but the key question is: how close can you get? The mathematicians of the time knew how to answer that; you can do it by calculus or even just algebra. But first, you need one extra ingredient: a definition of the total error. Legendre's first idea was to add the individual errors together, but that doesn't quite work. If the answer should be 5, and you get 4 and 6 on two different occasions, the individual errors are −1 and +1, which add to zero. Errors in opposite directions cancel. To avoid this, Legendre needed to convert all the errors to positive numbers.

One way to do this is to replace the errors by their absolute values, reversing the sign of the negative ones. Unfortunately this formulation leads to messy algebra with no neat answer (though today we can handle it using computers). Instead, he *squared* all the errors and then added them. The square of either a positive or a negative number is always positive, and squares lead to tractable algebra. Minimising the sum of the squares of the errors is an easy problem, and there's a simple formula for the solution. Legendre called this technique (in

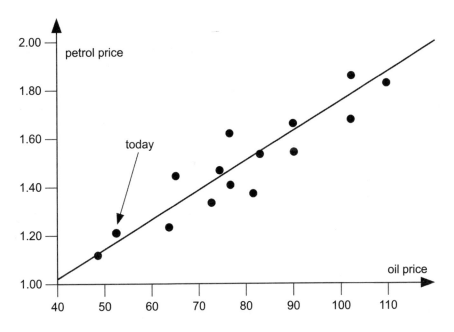

Hypothetical data on price of petrol against price of oil. Dots: data. Line: fitted linear relationship.

French) the *method of least squares*, saying that it is 'appropriate for revealing the state of the system which most nearly approximates the truth'.

A common application, mathematically identical when we're considering just two variables, is to fit a straight line through a set of numerical data relating those variables. For instance, how does the price of petrol relate to the price of oil? Maybe the prices today are $52·36 per barrel for oil and £1·21 per litre for petrol. That gives you one data point with coordinates (52·36, 1·21). Visit several petrol stations on several days and you get, say, 20 such pairs. (In real applications, it might be hundreds, or even millions.) Suppose you want to predict the future price of petrol from that of oil. You plot the oil price along the horizontal axis and that of petrol along the vertical axis, getting a scatter of dots. Even in this picture you can see a general trend: not surprisingly, the higher the oil price, the higher that of petrol. (However, sometimes oil goes down but petrol stays the same, or goes up. It never goes down when oil goes up, in my experience.

Increases in the oil price are passed immediately to the consumer; decreases take far longer to show up at the forecourt.)

With Legendre's mathematics, we can do better. We can find the straight line that passes through the cloud of points, and comes as close as possible to them, in the sense that the sum of the squares of the errors is made as small as possible. The equation of this line tells us how to predict the price of petrol from the price of oil. For example, it might tell us that the price of petrol (in pounds per litre) is best approximated by multiplying the oil price (in dollars per barrel) by 0·012 and adding 0·56. It won't be perfect, but it will make the error as small as possible. With more variables it's not very helpful to draw a diagram, but the identical mathematical technique gives the best possible answer.

Legendre's idea provides one very simple but very useful answer to the big question discussed in this book: *How do we deal with uncertainty?* His answer is: make it as small as we can.

Of course, it's not quite as easy as that. Legendre's answer is 'best possible' for that particular measure of the overall error, but with a different choice, a different straight line might minimise the error. Moreover, the method of least squares has some faults. The trick of squaring the errors simplifies the algebra beautifully, but it can give too much weight to 'outliers' – data points that are way out of line with all the rest. Maybe some garage charges £2·50 a litre when everyone else is charging £1·20. Squaring the error makes the effect of the garage's price gouging a lot bigger than it would otherwise have been, distorting the entire formula. A pragmatic solution is to throw away outliers. But sometimes, in science or in economics, outliers are important. If you throw them away, you miss the point. You could prove that everyone in the world is a billionaire if you threw away all data relating to everybody else.

Also, if you're permitted to throw data away, you can remove any observations that contradict what you're setting out to prove. Many scientific journals now require *all* experimental data related to any particular publication to be available online, so that anyone can check that there's not been any cheating. Not that they expect much of that, but it's good practice to be seen to be honest. And occasionally scientists do cheat, so it helps prevent dishonesty.

Legendre's method of least squares isn't the last word in the statistical analysis of relationships between data. But it was a good *first* word – the first truly systematic method for extracting meaningful results from overdetermined equations. Later generalisations allow more variables, fitting multidimensional 'hyperplanes' to data instead of lines. We can also work with *fewer* variables, in the following sense. Given a set of data, which single value is the best fit in a least-squares sense? For instance, if the data points are 2, 3, and 7, which single value makes the sum of the squares of differences from the data as small as possible? A quick bit of calculus shows that this value is the mean (another word for average) of the data, $(2 + 3 + 7)/3 = 4$.[16] Similar calculations show that the best-fitting estimate to any list of data, in a least-squares sense, is the mean.

NOW THE STORY TAKES A STEP back in time, to pick up a new thread. When the numbers get large, binomial coefficients are hard to calculate by hand, so the early pioneers looked for accurate approximations. The one we still use today was discovered by Abraham de Moivre. He was born in France in 1667, but fled to England in 1688 to escape religious persecution. In 1711 he started publishing on probability, and collected his thoughts in *Doctrine of Chances* in 1718. At that stage he despaired of applying Bernoulli's insights to areas such as economics and politics, because it's difficult to calculate the binomial distribution when the number of trials is large. But shortly after, he started to make progress by approximating the binomial distribution by something more tractable. He published his preliminary results in his *Miscellanea analytica* of 1730, and three years later he completed the task. In 1738 he incorporated the new ideas into *Doctrine of Chances*.

He began by approximating the largest binomial coefficient, the one in the middle, and discovered that it's roughly $2^{n+1}/\sqrt{2\pi n}$, where n is the number of trials. He then tried to deduce the values of the other binomial coefficients by working outwards from the middle. In 1733 he derived an approximate formula, relating Bernoulli's binomial distribution to what we now call the normal distribution. It was to prove fundamental to the development of both probability theory and statistics.

The normal distribution forms an elegant curve, with a single peak in the middle, like the binomial distribution that it approximates. It's

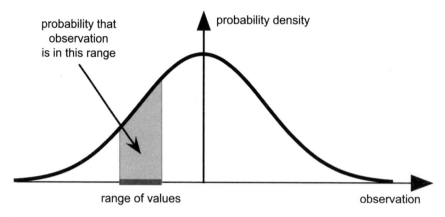

probability that
observation
is in this range

probability density

range of values

observation

The area under the normal distribution curve, between a given range of values, is the probability of an observation lying in that range.

symmetric about the peak, and tails off rapidly on either side, which allows the total area to be finite; indeed, equal to 1. This shape is vaguely like that of a bell, hence the alternative (modern) name 'bell curve'. It's a continuous distribution, which means that it can be evaluated at any real number (infinite decimal). Like all continuous distributions, the normal distribution doesn't tell us the probability of getting any *specific* measurement. That's zero. Instead, it tells us the probability that a measurement falls within some given range of values. This probability is the area under the part of the curve that lies within the range concerned.

Statisticians like to use an entire family of related curves, in which the axes are 'scaled' to change the mean and standard deviation. The formula is

$$\frac{1}{\sqrt{2\pi\sigma^2}}e^{-(x-\mu)^2/2\sigma^2}$$

which is denoted by $N(\mu, \sigma^2)$ for short. Here μ is the mean (another name for 'average') giving the location of the central peak, and σ is the standard deviation, a measure of the 'spread' of the curve – how wide its central region is. The mean tells us the average value of data consistent with the normal distribution, and the standard deviation tells us how big fluctuations about the mean are, on average. (The square of the standard deviation, σ^2, is called the *variance*, and this is

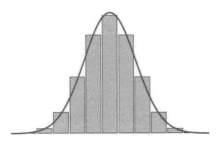 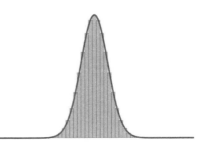

Left: Normal approximation to binomial distribution for ten trials, H and T equally likely. *Right*: The approximation is even closer for fifty trials.

sometimes simpler to work with.) The factor involving π makes the total area equal to 1. Its appearance in a probability problem is remarkable because π is usually defined in connection with circles, and it's not clear how circles relate to the normal distribution. Nevertheless, it's exactly the same number as the ratio of the circumference of a circle to its diameter.

De Moivre's big discovery, stated in these terms, is that for a large number n of trials, the bar chart for a binomial distribution has the same shape as the normal distribution $N(\mu, \sigma^2)$ with $\mu = n/2$ and $\sigma^2 = n/4$.[17] Even when n is small, it's a pretty good approximation. The left-hand figure shows the bar chart and the curve for ten trials, and the right-hand one does the same for fifty trials.

PIERRE-SIMON DE LAPLACE WAS another mathematician with strong interests in astronomy and probability. His masterwork *Traité de mécanique céleste* (Treatise on Celestial Mechanics), published in five volumes between 1799 and 1825, occupied much of his time, but after the fourth volume appeared in 1805 he took up an old idea and completed it. In 1810 he revealed what we now call the *central limit theorem* to the French Academy of Sciences. This far-reaching generalisation of de Moivre's result cemented the special role of the normal distribution in statistics and probability. Laplace proved that not only does the total number of successes in many trials approximate a normal distribution, as de Moivre had shown, but so does the total of *any* sequence of random variables drawn from the same probability distribution – *whatever that distribution might be*. The same goes for the mean, which is the total divided by the number of trials, but the scale on the horizontal axis must be adjusted accordingly.

Let me unpack that. An observation, in astronomy or elsewhere, is a number that can vary over some range because of errors. The associated 'error distribution' tells us how probable any given size of error is. However, we usually don't know what that distribution is. The central limit theorem says that this doesn't actually matter, as long as we repeat the observation many times and take the average. Each series of observations leads to one such average. If we repeat the whole process many times, the resulting list of averages has its own probability distribution. Laplace proved that this distribution is always approximately normal, and the approximation becomes as good as we wish if we combine enough observations. The unknown error distribution affects the mean and standard deviation of these averages, but not their overall pattern. In fact, the mean stays the same, and the standard deviation is divided by the square root of the number of observations in each average. The bigger that number is, the more tightly the averaged observations cluster around the mean.

This explains why Mayer's claim that combining nine times as many observations divides the error by 9 is wrong. Nine should be replaced by its square root, which is 3.

At much the same time, the great German mathematician Carl Friedrich Gauss used the same bell-shaped function when discussing the method of least squares in his epic astronomy text *Theoria motus corporum coelestium in sectionibus conicis solum ambientium* (Theory of the Motion of Heavenly Bodies Moving about the Sun in Conic Sections) of 1809. He used probabilities to motivate least squares, by seeking the most likely straight-line model of the data. To work out how likely a given line was, he needed a formula for the error curve: the probability distribution of observational errors. To obtain this, he assumed that the mean of many observations is the best estimate of the true value, and deduced that the error curve is a normal distribution. He then proved that maximising the likelihood leads to the standard least-squares formula.

This was a strange way to go about it. Stigler points out that the logic was circular, and had a gap. Gauss argued that the mean (a special case of least squares) is only 'most probable' if the errors are normally distributed; the mean is generally acknowledged as a good way to combine observations so that the error distribution is normal, so assuming a normal distribution leads back to least squares. Gauss criticised his own approach in later publications. But his result had instant resonance for Laplace.

Until that moment, Laplace hadn't even dreamed that his central limit theorem had any connection with finding best-fitting straight lines. Now he realised that it justified Gauss's approach. If observational errors result from the combination of many small ones – a reasonable assumption – then the central limit theorem implies that the error curve must be (approximately) normal. This in turn implies that the least-squares estimate is the best in a natural probabilistic sense. It's almost as if the pattern of errors is governed by a series of random coin tosses, so that each head makes the observation too big by a specific tiny amount, and each tail makes the observation too small by the same tiny amount. The whole story came together in one neat package.

In Chapter 4 we looked at the totals thrown by one, two, and three dice. I'll calculate the means and standard deviations of those distributions. The mean is in the middle. For one dice, the middle is between 3 and 4, at 3·5. For two the mean of the total is 7. For 3 it lies between 10 and 11, at 10·5. We can think of rolling two dice as two observations of the dice. The average *observation* is then the total divided by 2. If we divide 7 by 2 we get 3·5, the mean for one dice. The same goes for three dice: divide 10·5 by 3 and we again get 3·5. This shows that averaging observations from a given distribution leads to the same mean. The standard deviations are respectively 1·71, 1·21, and 0·99, with ratios 1 to $1/\sqrt{2}$ to $1/\sqrt{3}$, in accordance with the central limit theorem.

WHEN THE NORMAL DISTRIBUTION IS a good model of a probabilistic process, it lets us calculate the probability of measurements being in any specific range. In particular, we can find the probability that an observation deviates from the mean by some fixed amount by calculating the corresponding area under the normal curve. Because the width of the curve scales like the standard deviation σ, these results can be expressed in a form that applies to any normal distribution, whatever its standard deviation may be. The sums show that roughly

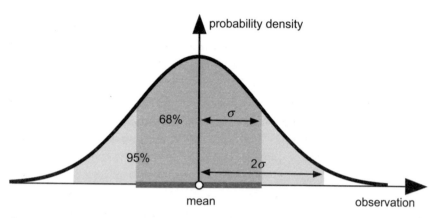

The mean (or average) of the normal curve is in the middle; the standard deviation σ measures the spread of observations around the mean. Events within σ of the mean occurs around 68% of the time; events within 2σ occur around 95% of the time.

68% of the probability lies within $\pm\sigma$ of the mean, and 95% lies within $\pm 2\sigma$ of the mean. These figures mean that the probability of an observation differing from the mean by more than σ is about 32%, while a difference of 2σ occurs only 5% of the time. As the difference gets bigger, the probability decreases very rapidly:

probability of differing from mean by more than σ is 31·7%
probability of differing from mean by more than 2σ is 4·5%
probability of differing from mean by more than 3σ is 2·6%
probability of differing from mean by more than 4σ is 0·006%
probability of differing from mean by more than 5σ is 0·00006%
probability of differing from mean by more than 6σ is 0·0000002%

In most biological and medical studies, the 2σ level is considered interesting and the 3σ level is decisive. The financial markets, in particular, characterise how unlikely some event, such as the price of a stock dropping by 10% in a few seconds, is, using phrases like 'a four-sigma event', meaning one that according to the normal distribution should occur only 0·006% of the time. We'll see in Chapter 13 that the normal distribution isn't always appropriate for financial data, where 'fat tails' can make extreme events far more common than a normal distribution would suggest. In particle physics, where the existence of a new fundamental particle depends on teasing out statistical evidence from millions of particle collisions, an important new discovery isn't considered publishable, or even worthy of being announced in the press, until the chance that it's a statistical accident passes the 5σ level. Roughly, the probability that the apparent detection of the particle is an accident is then one in a million. The detection of the Higgs boson in 2012 was not announced until the data reached that level of confidence, although the researchers used preliminary results at the 3σ level to narrow the range of energies to be examined.

WHAT, EXACTLY, *IS* PROBABILITY? So far, I've defined it rather vaguely: the probability of an event is the proportion of trials in which it happens, in the long run. The justification for this 'frequentist' interpretation is Bernoulli's law of large numbers. However, in any particular series of trials, that proportion fluctuates, and it's very seldom equal to the theoretical probability of the event.

We might try to define probability as the limit of this proportion as the number of trials gets large, in the sense of calculus. The limit of a sequence of numbers (if it exists) is the unique number such that given any error, however small, the numbers in the sequence differ from the limit by less than that error, if you go far enough along the sequence. The problem is that, very rarely, a series of trials might give a result like

H H H H H H H H H H H H H H H H H H H H H H ...

with heads turning up every time. Or there might be a scattering of Ts dominated by a lot more Hs. Agreed, such sequences are unlikely, but they're not impossible. Now: what do we mean by 'unlikely'? Very small probability. Apparently we need to define 'probability' to define the right kind of limit, and we need to define the right kind of limit to define 'probability'. It's a vicious circle.

Eventually mathematicians twigged that the way to get round this obstacle was to borrow a trick from the ancient Greek geometer Euclid. Stop worrying about what probabilities *are*; write down what they *do*. More precisely, write down what you want them to do – the general principles that had emerged from all that previous work. These principles are called axioms, and everything else is deduced from them. Then, if you want to apply probability theory to reality, you make the hypothesis that probabilities are involved in some specific manner. You use the axiomatic theory to work out the consequences of this assumption; then you compare the results with experiment, to see if the hypothesis is correct. As Bernoulli understood, his law of large numbers justifies using the observed frequencies to estimate probabilities.

It's fairly easy to set up axioms for probability when the number of events is finite, such as the two sides of a coin or the six sides of a dice. If we write $P(A)$ for the probability of event A, the main properties we need are positivity

$$P(A) \geqslant 0$$

the universal rule

$$P(U) = 1$$

where U is the universal event 'any outcome', and the addition rule

$$P(A \text{ or } B) = P(A) + P(B) - P(A \text{ and } B)$$

because A and B can overlap. If they don't, it simplifies to

$$P(A \text{ or } B) = P(A) + P(B)$$

Since A or not-$A = U$, we easily deduce the negation rule

$$P(\text{not}-A) = 1 - P(A)$$

If two independent events A and B occur in succession, we *define* the probability that they both occur to be

$$P(A \text{ and then } B) = P(A)P(B)$$

We can then prove that this probability satisfies the above rules. You can trace these rules right back to Cardano, and they're pretty much explicit in Bernoulli.

This is all very well, but the advent of continuous probability distributions, unavoidable thanks to de Moivre's brilliant breakthrough, complicates the axioms. This is unavoidable, because measurements need not be whole numbers, so applications of probability also demand continuous distributions. The angle between two stars, for instance, is a continuous variable, and it can be anything between 0° and 180°. The more accurately you can measure it, the more important a continuous distribution becomes.

There's a useful clue. When discussing the normal distribution, I said that it represents probabilities as areas. So we should really axiomatise the properties of areas, throwing in the rule that the total area under the curve is 1. The main extra rule for continuous distributions is that the addition formula works for infinitely many events:

$$P(A \text{ or } B \text{ or } C \text{ or} \ldots) = P(A) + P(B) + P(C) + \ldots$$

on the assumption that none of A, B, C, and so on, overlap. The ... indicates that both sides can be infinite; the sum on the right makes sense (converges) because every term is positive and the total never exceeds 1. This condition lets us use calculus to work out probabilities.

It's also useful to generalise 'area' to any quantity that behaves in the same manner. Volume in three dimensions is an example. The upshot of this approach is that probabilities correspond to 'measures', which assign something akin to area to suitable subsets (called 'measurable') of a space of events. Henri Lebesgue introduced measures in the theory of integration in 1901–2, and the Russian mathematician Andrei Kolmogorov used them to axiomatise probability in the 1930s, as follows. A *sample space* comprises a set, a collection of subsets called *events*, and a measure P on the events. The axioms state that P is a measure, and that the measure of the whole set is 1 (that is, the probability that *something* happens is 1). That's all you need, except that the collection of events must have some technical set-theoretic properties that let it support a measure. Exactly the same set-up applies to finite sets, but you don't need to fiddle around with infinities. Kolmogorov's axiomatic definition dispensed with several centuries of heated controversy, and gave mathematicians a well-defined notion of probability.

A more technical term for a sample space is *probability space*. When we apply probability theory in statistics, we model the sample space of possible real events as a probability space in Kolmogorov's sense. For example, when studying the ratio of boys to girls in a population, the real sample space is all children in that population. The model we compare this to is the probability space consisting of four events: the empty set $\emptyset$, G, B, and the universal set $\{G, B\}$. If boys and girls are equally likely, the probabilities are $P(\emptyset) = 0, P(G) = P(B) = 1/2, P(\{G, B\}) = 1.$

For simplicity, I'm going to use 'sample space' to refer both to real events and to theoretical models. The important thing is to make sure we choose an appropriate sample space, in either sense, as the puzzles in the next chapter demonstrate.

6

FALLACIES AND PARADOXES

I have seen men, ardently desirous of having a son, who could learn only with anxiety of the births of boys in the month when they expected to become fathers. Imagining that the ratio of these births to those of girls ought to be the same at the end of each month they judged that the boys already born would render more probable the births next of girls.

Pierre-Simon de Laplace, *A Philosophical Essay on Probabilities*

HUMAN INTUITION FOR PROBABILITY IS hopeless.

When we're asked to give a rapid estimate of the odds of chance events, we often get them completely wrong. We can train ourselves to improve, as professional gamblers and mathematicians have done, but that takes time and effort. When we make a snap judgement of how probable something is, likely as not we get it wrong.

I suggested in Chapter 2 that this happens because evolution favours 'quick and dirty' methods when a more reasoned response could be dangerous. Evolution prefers false positives to false negatives. When the choice is between a half-hidden brown thing being either a leopard or a rock, even a single false negative can prove deadly.

Classic probability paradoxes (in the sense of 'giving a surprising result', rather than self-contradictory logic) support this view. Consider the birthday paradox: how many people have to be in a room before it becomes more likely than not that two of them have the same birthday? Assume 365 days in the year (no 29 February dates), and all of them equally likely (which isn't quite true, but hey). Unless people have come across this one before, they go for quite large numbers: 100, perhaps, or 180 on the grounds that this is roughly half of 365. The correct answer is 23. If you want to know why, I've put the reasoning in the Notes at the back.[18] If the distribution of births isn't uniform, the answer can be smaller than 23, but not larger.[19]

Here's another puzzle that people often find confusing. The Smiths have exactly two children, and at least one is a girl. Assume for simplicity that boys and girls are equally likely (actually boys are slightly more likely, but this is a puzzle, not a scholarly paper on demography) and that children are one or the other (gender and unusual chromosome issues ignored). Assume also that the sexes of the children are independent random variables (true for most couples but not all). What is the probability that the Smiths have two girls? No, it's not 1/2. It's 1/3.

Now suppose that the *elder* child is a girl. What is the probability that they have two girls? This time, it really is 1/2.

Finally, suppose that at least one is a girl born on a Tuesday. What is the probability that they have two girls? (Assume all days of the week are equally likely – also not true in reality, but not too far off.) I'll leave that one for you to think about for a while.

The rest of this chapter examines some other examples of paradoxical conclusions and faulty reasoning in probability. I've included some old favourites, and some less well-known examples. Their main purposes is to drive home the message that when it comes to uncertainty, we need to think very carefully and not make snap judgements. Even when effective methods exist to deal with uncertainties, we have to be aware that they can mislead us if misused. The key concept here, conditional probability, is a running theme in the book.

A COMMON ERROR, WHEN FACED with a choice between two alternatives, is to make the natural default human assumption that the chances are even – fifty-fifty. We talk blithely of events being 'random', but we seldom examine what that means. We often equate it with things being just as likely to happen as not: fifty-fifty odds. Like tossing a fair coin. I made just this assumption at the end of the second paragraph of this chapter, when I wrote: 'likely as not'. Actually, what's likely, and what's not, seldom occur with equal probabilities. If you think about what the words in the phrase mean, this becomes obvious. 'Not likely' means low probability, 'likely' means higher. So even our standard phrase is confused.

Those time-honoured probability puzzles show that getting it wrong can be a lot more likely than not. In Chapter 8 we'll see that our poor understanding of probabilities can lead us astray when it really matters, such as deciding guilt or innocence in a court of law.

Here's a clear-cut case where we can do the sums. If you're presented with two cards lying on a table and are told (truthfully!) that one is the ace of spades, and the other isn't, it seems obvious that your chance of picking the ace of spades is 1/2. That's true for that scenario. But in very similar situations, the default fifty-fifty assumption that evolution has wired into our brains is completely wrong. The classic example is the Monty Hall problem, much beloved of probability theorists. It's almost a cliché nowadays, but some aspects often get overlooked. Additionally, it's a perfect route into the counterintuitive territory of *conditional* probability, which is where we're heading. This is the probability of some event occurring, *given that* some other event has already happened. And it's fair to say that when it comes to conditional probabilities, the default assumptions that evolution has equipped our brains with are woefully inadequate.

Monty Hall was the original host of the American TV game show *Let's Make a Deal*. In 1975 the biostatistician Steve Selvin published a paper on a version of the show's strategy; it was popularised (amid huge and largely misplaced controversy) by Marilyn vos Savant in her column in *Parade* magazine in 1990. The puzzle goes like this. You're presented with three closed doors. One conceals the star prize of a Ferrari; each of the others conceals the booby prize of a goat. You choose one door; when it's opened you'll win whatever is behind it. However, the host (who knows where the car is) then opens one of the other two doors, which he *knows* reveals a goat, and offers you the opportunity to change your mind. Assuming you prefer a Ferrari to a goat, what should you do?

This question is an exercise in modelling as well as in probability theory. A lot depends on whether the host always offers such a choice. Let's start with the simplest case: he always does, and everyone knows that. If so, you double your chances of winning the car if you switch.

This statement immediately conflicts with our default fifty-fifty brain. You're now looking at two doors. One hides a goat; the other a car. The odds must surely be fifty-fifty. However, they're not, because what the host did is conditional on which door you chose. Specifically, he didn't open *that* door. The probability that a door hides the car, *given that it's the one you chose*, is 1/3. That's because you chose from three doors, and the car is equally likely to be behind each of them because you had a free choice. In the long run, your choice will win the car one time in three – so it will not win the car two times in three.

Since another door has been eliminated, the conditional probability that the car is behind a door, *given that this is not the one you chose*, is $1 − 1/3 = 2/3$, because there's only one such door, and we've just seen that two times out of three your door is the wrong one. Therefore, two times out of three, changing to the other door wins the car. That's what Steve said, it's what Marilyn said, and a lot of her correspondents didn't believe it. But it's true, with the stated conditions on what the host does.

If you're sceptical, read on.

One psychological quirk is fascinating: people who argue that the odds must be fifty-fifty (so either door is equally likely to win the car) generally prefer *not* to change their minds, even though fifty-fifty odds imply that swapping won't do any harm. I suspect this is related to the modelling aspect of the problem, which involves the sneaking suspicion – very possibly correct – that the host is out to fool you. Or maybe it's the Bayesian brain, believing it's being fooled.

If we abandon the condition that the host always offers you the opportunity to change your mind, the calculation changes completely. At one extreme, suppose the host offers you the opportunity to change doors only when your choice would win the car. Conditional on him offering that choice, your door wins with probability 1, and the other one wins with probability 0. At the other extreme, if the host offers you the opportunity to change doors only when your choice would not win the car, the conditional probabilities are the other way round. It seems plausible that if the host mixes up these two situations in suitable proportions, your chance of winning by staying with your original choice can be anything, and so can your chance of losing. Calculations show that this conclusion is correct.

Another way to see that fifty-fifty can't be right is to generalise the problem and consider a more extreme example. A stage magician spreads out a pack of cards face down (an ordinary pack with 52 different cards, nothing rigged) and offers you a prize if you can pick the ace of spades. You choose a card and slide it out, *still* face down. The magician picks up the other 51 cards, looks at them all while hiding them from you, and starts placing them face up on the table, none being the ace of spades. He keeps going for some time, then puts one card face down next to yours, then resumes placing cards that aren't the ace of spades face up until the whole pack has been dealt with. Now there are 50 cards face up, none the ace of spades, and two face down: the one you chose initially, and the one he placed next to it.

Assuming he didn't cheat – which is perhaps silly if he's a stage magician, but I assure you that on this occasion he didn't – which card is more likely to be the ace of spades? Are they equally likely? Hardly. Your card was chosen at random from the 52 in the pack, so it will be the ace of spades one time out of 52. The ace of spades will be in the rest of the pack 51 times out of 52. If it is, the magician's card must be the ace of spades. On the rare occasion that it's not – one time out of 52 – your card is the ace of spades, and the magician's card is whatever was left after he discarded 50 cards. So the probability your card is the ace of spades is 1/52; the probability the magician's card is the ace of spades is 51/52.

The fifty-fifty scenario does arise, however, under suitable conditions. If someone who has not seen what has transpired is brought on stage, shown those two cards, and asked to pick which of them is the ace of spades, their chance of succeeding is 1/2. The difference is that you chose your card at the start of the proceedings, and what the magician did was conditional on that choice. The newcomer turns up after the process is complete, so the magician can't do anything conditional on their choice.

To drive the point home, suppose we repeat the procedure but this time you turn your card *face up* before the magician starts discarding cards. If your card isn't the ace of spades, his card has to be (again assuming no sleight of hand). And this happens 51 times out of 52 in the long run. If your card is the ace of spades, then his can't be; this happens one time out of 52 in the long run.

The same argument works with the car and goats if you *open the door you chose*. One time in three you'll see the car. The other two times, you'll see two open doors with goats and one closed door. *Where do you think the car is?*

BACK TO THE SMITHS AND their children, a simpler puzzle but just as deceptive. Recall the two versions:

1 The Smiths have exactly two children, and you're told that at least one is a girl. Assuming boys and girls are equally likely, along with the other conditions I mentioned, what is the probability that they have two girls?

2 Now suppose you're told that the *elder* child is a girl. What is the probability that they have two girls?

The default reaction to the first question is to think: 'One is a girl. The other one is equally likely to be a boy or a girl.' That leads to the answer 1/2. The flaw is that they might have two girls (after all, that's the event whose chances we're being asked to estimate), in which case '*the* other one' isn't uniquely defined. Think of the two births happening in order (even with twins, one is born first). The possibilities are

GG GB BG BB

We assumed the sex of the second child is independent of that of the first, so these four possibilities are equally likely. If all cases can occur, each has probability 1/4. However, the extra information rules out BB. We're left with three cases, and they're still equally likely. Only one of them is GG, so the probability of that is 1/3.

It looks as though the probabilities here have changed. Initially, GG has probability 1/4; suddenly it's 1/3. How come?

What's changed is the context. These puzzles are all about the appropriate *sample spaces*. The extra information 'not BB' cuts the sample space down from four possibilities to three. The real-world sample space now consists not of all families with two children, but of families with two children, not both boys. The corresponding model sample space consists of GG, GB, BG. These are all equally likely, so their probability in that sample space is 1/3, not 1/4. BB is irrelevant because in this case it can't occur.

It's not paradoxical that extra information changes the relevant probabilities. If you're betting on Galloping Girolamo, and you get a hot tip that the favourite Barnstorming Bernoulli has some mystery

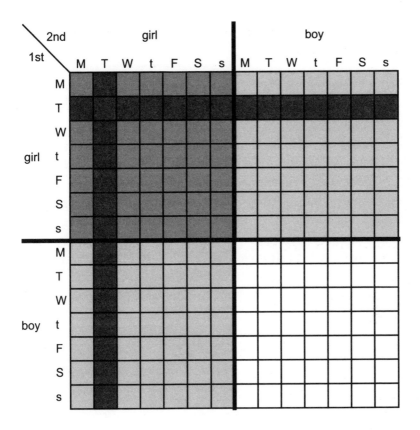

Sample spaces for the Tuesday girl. Grey shading: 'at least one girl'. Medium grey shading: 'both are girls'. Dark grey shading: 'at least one girl born on Tuesday'.

illness that slows it down, your chance of winning has definitely increased.

This puzzle is another example of conditional probability. Mathematically, the way to calculate a conditional probability is to cut the sample space down by including only the events that still might occur. In order to make the total probability for the smaller sample space equal to 1, the previous probabilities must all be multiplied by a suitable constant. We'll see *which* constant shortly.

In the third version of the puzzle, we know that at least one child is a girl born on a Tuesday. As before, the question is: What is the probability that the Smiths have two girls? I'll call this the target event. The probability we want is the chance that the Smiths hit the target. As

I said, we'll assume all days of the week are equally likely, to keep the account simple.

At first sight, the new information seems irrelevant. What does it matter which day she's born on? They're all equally likely! But before leaping to conclusions, let's look at the appropriate sample spaces. The picture shows all possible combinations of sex and day for both the first and second child. This is the full sample space, and each of its 196 squares (14 × 14) is equally likely, probability 1/196. The top left-hand quadrant, shaded light grey except where overlapped by a dark square, contains 49 squares. This corresponds to the event 'both are girls'; the probability of that on its own is 49/196 = 1/4, as expected.

The new information 'at least one is a girl born on a Tuesday' cuts the sample space down to the two dark stripes. In total these contain 27 squares: 14 horizontal, plus 14 vertical, minus 1 for the overlap because we mustn't count the same event twice. In our new cut-down sample space, these events are still all equally likely, so the conditional probability of each is 1/27. Count how many dark squares lie in the 'both girls' target region: the answer is 13 (7 + 7, minus 1 for the overlap). The other 14 lie in regions where the Smiths have at least one boy, so they miss the target. All small squares are equally likely, so the conditional probability that the Smiths have two girls, given that at least one is a girl born on Tuesday, is 13/27.

The birth day *does* matter!

I doubt anyone would be likely to guess this answer, except by sheer accident, unless they're a statistician who's good at mental arithmetic. You have to do the sums.

However, if instead we'd been told that at least one child is a girl born on Wednesday, or Friday, we'd have got the same conditional probability, using different stripes in the picture. In that sense, the day *doesn't* matter. So what's going on?

Sometimes telling people a counterintuitive piece of mathematics leads them to conclude that the mathematics is useless, not to embrace its surprising power. There's a danger of this occurring here, because some people instinctively reject this answer. It just doesn't make sense to them that the day she's born can change the probabilities. The calculation alone doesn't help much if you feel that way; you strongly suspect that there's a mistake. So some kind of intuitive explanation is needed to reinforce the sums.

The underlying error in the reasoning 'the day she's born can't change anything' is subtle but crucial. The choice of day is irrelevant, but picking some specific day does matter, because there might not be a specific *she*. For all we know – indeed, this is the point of the puzzle – the Smiths might have two girls. If so, we know that *one* of them is born on Tuesday, but not which one. The two simpler puzzles show that extra information that improves the chance of distinguishing between the two children, such as which one is born first, changes the conditional probability of two girls. If the elder is a girl, that probability is what we expect: 1/2. (The same would also be true if the younger were a girl.) But if we don't know which child is the girl, the conditional probability decreases to 1/3.

These two simpler puzzles illustrate the importance of extra information, but the precise effect isn't terribly intuitive. In the current version, it's not clear that the extra information does distinguish the children: we don't know *which* child was born on Tuesday. To see what happens, we count squares in the diagram.

The grid has three grey quadrants corresponding to 'at least one girl'. The medium-grey quadrant corresponds to 'both are girls', the pale-grey quadrants to 'the elder is a girl' and 'the younger is a girl', and the white quadrant corresponds to 'both are boys'. Each quadrant contains 49 smaller squares.

The information 'at least one girl' eliminates the white quadrant. If that's all we know, the target event 'both girls' occupies 49 squares out of 147, a probability of $49/147 = 1/3$. However, if we have the extra information 'the elder is a girl' then the sample space comprises only the top two quadrants, with 98 squares. Now the target event has probability $49/98 = 1/2$ These are the numbers I got before.

In these cases, the extra information increases the conditional probability of two girls. It does so because it reduces the sample space, but also because the extra information is consistent with the target event. This is the medium-grey region, and it lies inside both of the cut-down sample spaces. So the proportion of the sample space that it occupies goes up when the size of the sample space goes down.

The proportion can also go down. If the extra information was 'the elder child is a boy', the sample space becomes the bottom two quadrants, and the entire target event has been eliminated: its conditional probability decreases to 0. But whenever the extra information is consistent with the target event, it makes that event more likely, as measured by conditional probability.

The more *specific* the extra information is, the smaller the sample space becomes. However, depending on what that information is, it can also reduce the size of the target event. The outcome is decided by the interplay between these two effects: the first increases the conditional probability of the target, but the second decreases it. The general rule is simple:

conditional probability of hitting target, *given* information

$$= \frac{\text{probability of hitting target } \textit{and} \text{ being consistent with information}}{\text{probability of information}}$$

In the complicated version of the puzzle, the new information is 'at least one child is a girl born on Tuesday'. This is neither consistent with the target event nor inconsistent. Some dark-grey regions are in the top left quadrant, some are not. So we have to do the sums. The sample space is cut down to 27 squares, of which 13 hit the target and the other 14 don't. The overall effect is a conditional probability of 13/27, which is a good bit smaller than the 1/3 that we get without the extra information.

Let's check this result is consistent with the rule I just stated. 'Extra information' occurs in 27 squares out of 196, probability 27/196. 'Hitting target and agreeing with information' occurs in 13 squares out of 196, probability 13/196. My rule says that the conditional probability we want is

$$\frac{13/196}{27/196} = \frac{13}{27}$$

which is what we got by square counting. The 196s cancel, so the rule just expresses the square-counting procedure in terms of probabilities defined on the full sample space.

Notice that 13/27 is close to 1/2, which is what we would have got if we'd been told that the elder child is a girl. And this brings me back to the main point, and the reason why the conditional probability changes. Because it's possible for both children to be girls, it makes a big difference whether what we know is likely to distinguish between them. This is why 'one is a girl born on Tuesday' matters: the ambiguity when this information is true of *both* children has less effect. Why? Because even when the other child is a girl too, most of the time she will be born on a different day. Only 1/7 of the time is she born the same day. As we increase the chance of distinguishing between the two children, we move away from the 1/3 case (no distinction) to the 1/2 (we know exactly which one we're talking about).

The reason the answer here isn't exactly 1/2 is that in the target region, the two dark stripes, each with 7 squares, overlap by one square. Outside the target, the other two stripes don't overlap. So we get 13 stripes inside, 14 outside. The smaller this overlap, the closer the conditional probability gets to 1/2.

So here's a final version of the puzzle for you. All conditions as before, except that instead of the day of the week, we're told that one child is a girl, born on Christmas day. Assume all days of the year are equally likely (as always, not true in the real world) and that 29 February never happens (ditto). What is the conditional probability that both children are girls?

Would you believe 729/1459? See the Notes for the sums.[20]

Is this kind of hair-splitting about conditional probabilities important? In puzzles, no, unless you're a puzzle fan. In the real world, it can literally be a matter of life and death. We'll see why in Chapters 8 and 12.

IN EVERYDAY LANGUAGE, PEOPLE OFTEN talk of the 'law of averages'. This phrase may have arisen as a simplified statement of Bernoulli's law of large numbers, but in everyday usage it amounts to a dangerous fallacy, which is why you won't find mathematicians or statisticians using it. Let's see what's involved, and why they don't like it.

Suppose you toss a fair coin repeatedly, and keep a running count of how many times H and T turn up. There's a definite possibility of random fluctuations, so that at some stage the cumulative totals might give different numbers of heads and tails – say 50 more Hs than Ts. Part of the intuition behind the law of averages is that such an excess of heads ought to disappear if you keep going. That's correct, if suitably interpreted, but even so the situation is delicate. The mistake is to imagine that this excess of heads makes tails become more likely.

However, it's not totally unreasonable to imagine that they must; after all, how else can the proportions eventually balance out?

This kind of belief is encouraged by tables showing how often any particular number has come up in a lottery. Data for the UK's National Lottery can be found online. They're complicated by a change that increased the range of numbers that might be drawn. Between November 1994 and October 2015, when there were 49 numbers, the lottery machine spat out a ball bearing the number 12 on 252 occasions, whereas 13 turned up only 215 times. That was the least frequent number, in fact. The most frequent was 23, which was drawn 282 times. These results are open to many interpretations. Are the lottery machines unfair, so that some numbers are more likely to occur than others? Does 13 come up less often because, as we all know, it's unlucky? Or should we bet on 13 in future because it's got behind, and the law of averages says it has to catch up?

It's mildly curious that the worst number is 13, but whoever writes the script for the universe has a habit of using clichés. As it happens, 20 was also drawn 215 times, and I don't know of any superstitions about that number. Statistical analyses, based on Bernoulli's original principles, show that fluctuations of this magnitude are to be expected when the machines draw each number with equal probability. So there's no scientific reason to conclude that the machines aren't fair. Moreover, it's hard to see how the machine can 'know' what number is written on any particular ball, in the sense that the numbers don't influence the mechanics. The simple and obvious probability model of 49 equally likely numbers is almost certainly applicable, and the probability of 13 being drawn in the future is not influenced by what's happened in the past. It's no more or less likely than any other number, even if it's been a bit neglected.

The same goes for coins, for the same reason: if the coin is fair, a temporary excess of heads does not make tails become more likely. The probability of heads or tails is still 1/2. Above, I asked a rhetorical question: How else can the proportions balance out? The answer is that there's another way that can happen. Although an excess of heads has no effect on the subsequent *probability* of getting a tail, the law of large numbers implies that in the long term, the numbers of heads and tails do tend to even out. But it doesn't imply they have to become equal; just that their *ratio* gets closer to 1.

Suppose that initially we tossed 1000 times, getting 525 heads and 475 tails: an excess of 50 heads, and a ratio of $525/475 = 1·105$. Now suppose we toss the coin another two million times. On average we expect about one million heads and an equal number of tails. Suppose this happens, exactly. Now the cumulative scores are 1,000,525 heads and 1,000,475 tails. There are *still* 50 more heads. However, the ratio is now 1,000,525/1,000,475, which is 1·00005. This is much closer to 1.

AT THIS POINT, I'M FORCED to admit that probability theory tells us something stronger, and it sounds just like what people think of as the law of averages. Namely, whatever the initial imbalance may be, if you keep tossing for long enough then the probability that *at some stage* tails will catch up, and give exactly the same number as heads, is 1. Essentially, this is certain, but since we're talking of a potentially infinite process it's better to say 'almost certain'. Even if heads get a million throws ahead, tails will almost certainly catch up. You just have to keep tossing for long enough – though it will be very long indeed.

Mathematicians often visualise this process as a random walk. Imagine a pointer moving along the number line (positive and negative integers in order), starting at 0. At each toss of the coin, move it one step to the right if it's heads, one step to the left if it's tails. Then the position at any stage tells us the excess of Hs over Ts. For instance, if the tosses start HH it ends up two steps to the right at the number 2; if it's HT it goes one step right, one left, and ends up back at 0. If we graph the number it sits on against time, so left/right becomes down/up, we get a random-looking zigzag curve. For instance, the sequence

TTTTHTHHHHHHTTTHTTTH

(which I got by actually tossing a coin) gives the picture shown. There are 11 Ts and 9 Hs.

The mathematics of random walks tells us that the probability that the pointer *never* returns to zero is 0. Therefore the probability that eventually the numbers even out again is 1 – almost certain. But the theory also tells us some more surprising things. First, these statements

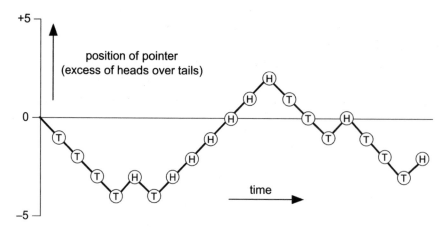

First 20 steps of a typical random walk.

are true even if we give H or T a big head start. Whatever the initial imbalance may be, it will almost certainly disappear if we keep tossing. However, the average time required for this to happen is *infinite*. That may seem paradoxical, but what it means is this. If we wait for the first return to 0, it takes some particular time. Keep going. Eventually it almost certainly makes a second return. This might take fewer steps than the first one, or it might take more. Every so often, it's a lot longer; indeed, if you choose any very large number you want, then almost certainly some return will take at least that long. If you average infinitely many arbitrarily large numbers, it's reasonable that you can get an infinite average.

This habit of repeatedly returning to 0 looks as if it conflicts with my statement that a coin has no memory. However, it doesn't. The reason is that despite everything I've said, there's also a sense in which coin tosses do *not* tend to even out in the long run. We've seen that the cumulative total is almost certain to be as large (negative or positive) as we wish, if we wait long enough. By the same token, it must eventually cancel out any initial imbalance.

Anyway, back to that tendency to even out if we wait long enough. Doesn't that prove the law of averages? No, because random walk theory has no implications for the probability of H or T turning up. Yes, they even out 'in the long run' – but we have no idea exactly how long that will be in any particular instance. If we stop at that precise moment, it looks like the law of averages is true. But that's cheating:

we stopped when we reached the result we wanted. Most of the time, the proportions weren't in balance. If we specify a fixed number of tosses in advance, there's no reason for heads and tails to even out after that number. In fact, on average, after any specific number of tosses, the discrepancy will be exactly the same as it was at the beginning.

7

SOCIAL PHYSICS

Let us apply to the political and moral sciences the method founded in observation
and calculation, the method that has served us well in the natural sciences.
Pierre-Simon de Laplace, *A Philosophical Essay on Probabilities*

IN ISAAC ASIMOV'S CLASSIC SCIENCE fiction novel *Foundation*, which
appeared in magazines in the 1940s and as a book in 1951, the
mathematician Hari Seldon forecasts the collapse of the Galactic
Empire using psychohistory: a calculus of the patterns that occur in the
reaction of the mass of humanity to social and economic events.
Initially put on trial for treason, on the grounds that his prediction
encourages said collapse, Seldon is permitted to set up a research group
on a secluded planet to minimise the destruction and reduce the
subsequent period of anarchy from 30,000 years to a mere thousand.

Asimov, like his readers, knew that predicting large-scale political
events over periods of millennia isn't really plausible, but that's a
matter of 'suspension of disbelief'. We all do it when reading fiction.
No Jane Austen fan gets upset to be told that Elizabeth Bennet and Mr
Darcy didn't actually exist. But Asimov was smart enough to know
that this kind of forecasting, however accurate it might be, is
vulnerable to any large disturbance that hasn't been anticipated, not
even in principle – to a 'black swan event', to use the popular
buzzword. He also understood that readers who happily swallowed
psychohistory would realise the same thing. So in the second volume of
the series, just such an event derails Seldon's plans. However, Seldon is
smart enough to plan for his plans going wrong, and he has a hidden
contingency plan, revealed in the third volume. It is also not what it
appears to be, yet another level of forward planning.

The *Foundation* series is notable for concentrating on the political
machinations of the key groups, instead of churning out page upon

page of space battles between vast fleets armed to the teeth. The protagonists receive regular reports of such battles, but the description is about as far from a Hollywood treatment as you can get. The plot is (as Asimov himself stated) modelled on Edward Gibbon's *History of the Decline and Fall of the Roman Empire*. The series is a masterclass in planning for uncertainty on an epic scale. Every senior minister and civil servant should be obliged to read it.

Psychohistory takes a hypothetical mathematical technique to extremes, for dramatic effect, but we use the basic idea every day for less ambitious tasks. Hari Seldon was to some extent inspired by a 19th-century mathematician, one of the first to take a serious interest in the application of mathematics to human behaviour. His name was Adolphe Quetelet, and he was born in 1796 in the Belgian city of Ghent. Today's obsessions with the promise (and dangers) of 'big data' and artificial intelligence are direct descendants of Quetelet's brainchild.

He didn't call it psychohistory, of course. He called it social physics.

THE BASIC TOOLS AND TECHNIQUES of statistics were born in the physical sciences, especially astronomy, as a systematic method to extract the maximum amount of useful information from observations that are subject to unavoidable errors. But as understanding of probability theory grew, and scientists started to become comfortable with this new method of data analysis, a few pioneers began to extend the method beyond its original boundaries. The problem of making the most accurate possible inferences from unreliable data arises in all areas of human activity. It is, quite simply, the search for maximum certainty in an uncertain world. As such, it has a particular appeal for any person or organisation needing to make plans now for events arising in the future. Which, to be honest, is virtually everyone; but, in particular, governments (national and local), businesses, and the military.

Within a relatively short period of time, statistics escaped the confines of astronomy and frontier mathematics, in an explosion of activity that has made it indispensable in all areas of science (especially the life sciences), medicine, government, the humanities, even the arts.

So it's fitting that the person who lit the fuse was a pure mathematician turned astronomer, who succumbed to the siren song of the social sciences and applied statistical reasoning to human attributes and behaviour. Quetelet bequeathed to posterity the realisation that despite all the vagaries of free will and circumstance, the behaviour of humanity in bulk is far more predictable than we like to imagine. Not perfectly, by any means; not totally reliably; but often 'good enough for government work', as they say.

He also bequeathed two more specific ideas, both hugely influential: *l'homme moyen*, or 'the average man', and the ubiquity of the normal distribution.[21] Both bequests have serious flaws if taken too literally or applied too widely, but they opened up new ways of thinking. Both are still alive today, flaws notwithstanding. Their main value is as 'proof of concept' – the idea that mathematics can tell us something significant about the way we act. This claim is controversial today (what isn't?), but it was even more so when Quetelet made his first tentative steps towards a statistical survey of human foibles.

Quetelet took a degree in science, the first doctorate awarded by the newly founded University of Ghent. His thesis was about conic sections, a topic going back to the ancient Greek geometers, who constructed important curves – ellipse, parabola, hyperbola – by slicing a cone with a plane. For a time he taught mathematics, until his election to the Royal Academy of Brussels propelled him into fifty-year career in the scholarly stratosphere, as the central figure of Belgian science. Around 1820, he joined a movement to found a new observatory. He didn't know much astronomy, but he was a born entrepreneur and he knew his way around the labyrinths of government. So his first step was to enlist the government's support and secure a promise of funding.

Only then did he take steps to remedy his ignorance of the subject that the observatory was intended to illuminate. In 1823, at government expense, he headed for Paris to study with leading astronomers, meteorologists, and mathematicians. He learned astronomy and meteorology from François Arago and Alexis Bouvard, and probability theory from Joseph Fourier and perhaps a by now elderly Laplace. That sparked a lifelong obsession with the application of probability to statistical data. By 1826, Quetelet was a regional correspondent for the statistical bureau of the Kingdom of the

Low Countries (Pays-Bas, today's Belgium and Holland). I'll say 'Belgian' rather than 'of the Low Countries' from now on.

IT ALL BEGAN INNOCENTLY ENOUGH.

One very basic number has a strong effect on everything that happens, and will happen, in a country: its population. If you don't know how many people you've got, it's difficult to plan anything sensibly. Of course you can guesstimate, and contingency plan for some degree of error, but that's all a bit rule-of-thumb. You may end up wasting a lot of money on unnecessary infrastructure, or underestimating demand and causing a crisis. This isn't just a 19th-century problem. Every nation grapples with it today.

The natural way to find out how many people live in your country is to count them. That is, carry out a census, which is not as easy as it might seem. People move around, and they hide themselves away to avoid being convicted of crimes or to avoid paying tax, or just to keep the government's prying nose out of what they fondly imagine to be their own private affairs. Anyway, in 1829 the Belgian government was planning a new census. Quetelet had been working for some time on historical population figures, and he got involved. 'The data that we have at present can only be considered provisional, and are in need of correction,' he wrote. The figures were based on older numbers, obtained under difficult political conditions; they had then been updated by adding the number of registered births and subtracting the number of registered deaths. This was a bit like navigating by 'dead reckoning': as time passed, errors would accumulate. And it missed out immigration/emigration altogether.

A full census is expensive, so it makes sense to use calculations along these lines to estimate the population between censuses. But you can't get away with it for very long. A census every ten years is common. Quetelet therefore urged the government to carry out a new census, to get an accurate baseline for future estimates. However, he'd come back from Paris with an interesting idea, which he'd got from Laplace. If it worked, it would save a lot of money.

Laplace had calculated the population of France by multiplying together two numbers. The first was the number of births in the past year. This could be found from the registers of births, which were pretty accurate. The other was the ratio of the total population to the annual number of births – the reciprocal of the birth rate. Clearly multiplying these together gives the total population, but it looks as though you need to know the total population to find the second number. Laplace's bright idea was that you could get a reasonable estimate by what we now call sampling. Select a smallish number of reasonably typical areas, perform a full census in those, and compare with the number of births in those areas. Laplace reckoned that about thirty such areas would be adequate to estimate the population of the whole of France, and did some sums to justify the claim.

In the event, however, the Belgian government eschewed sampling and carried out a full census. The reason for Quetelet's U-turn seems to have been an intelligent, informed, but woefully misguided criticism by Baron de Keverberg, an adviser to the state. Observing correctly that birth rates in different regions depend on a bewildering variety of factors, few of which can be anticipated, the baron argued that it would be impossible to create a representative sample. The errors would accumulate, making the results useless. In this, of course, he made the same mistake that Euler had made: assuming the worst case rather than the typical case. In practice, most sampling errors would have cancelled each other out through random variation. Still, it was a forgivable blunder, because Laplace had assumed that the best way to sample a population was to select, in advance, regions considered to be in some sense *representative* of the whole, with a similar mix of rich and poor, educated and uneducated, male and female, and so on.

Today, opinion polls are often set up along those lines, in an effort to get good results from small samples. It's an arcane business, and methods that used to work well seem to be failing ever more frequently, I suspect because everyone is fed up to the eyeballs with polls, market research questionnaires, and other unwanted intrusions. As statisticians eventually discovered, random samples are usually representative *enough* provided they're reasonably big. How big, we'll see later in this chapter. But all this was in the future, and Belgium duly tried to count every single person.

BARON DE KEVERBERG'S CRITICISM DID have one useful effect: it encouraged Quetelet to collect vast amounts of data in very precise circumstances, and analyse them to death. He soon branched out from counting people to *measuring* people, and comparing the measurements with other factors – season, temperature, geographical location. For eight years he collected data on birth rates, death rates, marriage, date of conception, height, weight, strength, growth rate, drunkenness, insanity, suicide, crime. He enquired into their variation with age, sex, profession, location, time of year, being in prison, being in hospital. He always compared just two factors at a time, which let him draw graphs to illustrate relationships. He put together a huge body of evidence regarding the quantitative nature of the variation of all of these variables across a typical population. In 1835 he published his conclusions in *Sur l'homme et le développement de ses facultés, ou Essai de physique sociale*, appearing in English in 1842 as *Treatise on Man and the Development of His Faculties*.

Significantly, whenever he referred to the book, he used the subtitle *'Physique sociale'*. And when he prepared a new edition in 1869, he swapped his former title and subtitle. He knew what he had created: a mathematical analysis of what it is to be human. Or, to avoid claiming too much, those features of humanity that can be quantified. One concept in the book caught the public imagination, and still does: that of the *average man*.

As a biologist friend of mine often remarked, the 'average man' has one breast and one testicle. In these gender-aware times, we must be very careful with terminology. Actually, Quetelet was perfectly well aware that – inasmuch as his concept made any sense at all – it was also necessary to consider the average woman, average child, indeed many different instances of all of these, for different populations. He noticed early on that his data for attributes such as height or weight (duly restricted to a single gender and age group) tended to cluster around a single value. If we draw the data as a bar chart or histogram, the tallest bar is in the middle, and the others slope away from it on either side. The whole shape is roughly symmetric, so the central peak – the commonest value – is also the average value.

I hasten to add that these statements are not exact, and are not applicable to all data, even human data. The distribution of wealth, for example, has a very different shape: the majority of people are poor, and a very small number of super-rich own half the planet. The

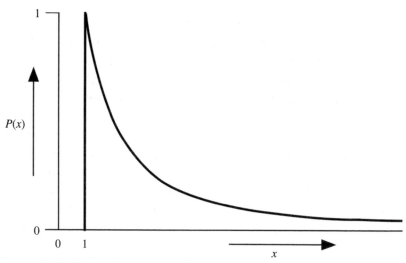

Pareto distribution based on a power x^a for some constant a, with a cut-off at $x = 1$.

standard mathematical model is a Pareto (power-law) distribution. But as an empirical observation, many types of data show this pattern, and it was Quetelet who realised its importance in social science. The general shape he found is, of course, the bell curve – the normal distribution of Euler and Gauss – or something sufficiently close to make that a reasonable mathematical model.

Extensive tables and graphs are all very well, but Quetelet wanted a snappy summary, one that conveyed the main point in a vivid, memorable manner. So instead of saying 'the average value of the bell curve for the heights of some class of human males over 20 years of age is 1·74', say, he came up with 'the average man in that class is 1·74 metres tall'. He could then compare these average men across different populations. How does the average Belgian infantryman stack up to the average French farmer? Is 'he' shorter, taller, lighter, heavier, or much the same? How does 'he' compare to the average German military officer? How does the average man in Brussels compare to his counterpart in London? What about the average woman? Average child? Average cat or dog, if you wish, though Quetelet stuck to people. Given all that, which country's average man is more likely to be a murderer or a victim? A doctor, devoted to saving lives, or a suicide, intent on ending his own?

The important point here is that Quetelet does *not* mean that a typical person, chosen at random, will match the average in all respects. Indeed, there's a sense in which that's impossible. Height and weight are approximately related in a manner that implies both can't be average at the same time. All else being equal, weight scales like volume, which scales like the cube of height. In a population of three cubes, with heights 1, 2, and 3 metres, the average height is 2. Their volumes are 1, 8, and 27, so the average volume is 12. Therefore the volume of the average cube is not the cube of the height of the average cube. In short, the average cube isn't a cube. In practice, this criticism is weaker than it looks, because human data are mostly concentrated near the mean. Cubes of height 1·9, 2, and 2·1 metres have mean height 2, and mean volume 8·04. Now the average cube is distinctly cube-ish.

Quetelet knew all this. He envisaged a different 'average man' (woman, child) for each attribute. To him, it was just a convenient form of words to simplify complicated statements. As Stephen Stigler put it: 'The average man was a device for smoothing away the random variations of society and revealing the regularities that were to be the laws of his "social physics".'[22]

SLOWLY, THE SOCIAL SCIENTISTS CAME to an awareness similar to that of the astronomers: valid inferences can be made by combining data from several sources, without having full knowledge or control of the different circumstances and magnitudes of likely errors. A certain amount of knowledge is needed, and better data generally provide more accurate answers, but the data themselves contain clues to the quality of the results.

After 1880, the social sciences began to make extensive use of statistical ideas, especially the bell curve, often as a substitute for experiments. A key figure is Francis Galton, who also pioneered data analysis in weather forecasting, discovering the existence of anticyclones. He produced the first weather map, published in the *Times* newspaper in 1875, and was fascinated by real-world numerical data and the mathematical patterns hidden within them. When Darwin published *The Origin of Species*, Galton began a study of human heredity. How does the height of a child relate to that of its parents? What about weight, or intellectual ability? He adopted Quetelet's bell curve, using it to separate distinct populations. If certain data showed two peaks, rather than the single peak of the bell curve, Galton argued that the population concerned must be composed of two distinct subpopulations, each following its own bell curve.[23]

Galton became convinced that desirable human traits are hereditary, a deduction from evolutionary theory that Darwin repudiated. For Galton, Quetelet's average man *was* a social imperative – one to be avoided. The human race should be more ambitious. His 1869 book *Hereditary Genius* invoked statistics to study the inheritance of genius and greatness, with what today is a curious mixture of egalitarian aims ('every lad [should have] a chance of showing his abilities, and, if highly gifted, [should be] enabled to achieve a first-class education and entrance into professional life') and the encouragement of 'the pride of race'. He coined the term 'eugenics' in his 1883 *Inquiries into Human Faculty and Its Development*, advocating financial rewards to encourage marriage between families of high rank, and the deliberate breeding of people with allegedly superior abilities. Eugenics had its day in the 1920s and 1930s, but fell rapidly from grace because of widespread abuses, such as forced sterilisation of mental patients and the Nazi delusion of a master race. Today it's generally considered racist, and it contravenes the United Nations Convention on the Prevention and Punishment of the Crime of Genocide, and the European Union's Charter of Fundamental Rights.

Whatever we think of Galton's character, his contribution to statistics was considerable. By 1877 his investigations were leading him to invent regression analysis, a generalisation of the method of least squares that compares one data set to another to find the most probable relationship. The 'regression line' is the best-fitting straight-line model of the relationship between the data, according to this method.[24] This led to another central concept in statistics, *correlation*, which quantifies the degree of relationship between two (or more) sets of data – for example, level of smoking and incidence of lung cancer. Statistical measures for correlation go back to the physicist Auguste Bravais, best known for his work on crystals. Galton discussed examples, such as the relation between forearm length and height in 1888.

Suppose you want to quantify how closely related human height and arm length are. You take a sample of individuals, measure these quantities, and plot the corresponding pairs of numbers. You then fit a straight line to these points using least squares, just like I fitted the price of petrol to that of oil. This method always gives *some* line, however scattered the data points are. Correlation quantifies how closely the line fits the cloud of data. If the data points are very close to the line, the two variables are highly correlated. If they spread out rather fuzzily around it, the correlation is less. Finally, if the line has negative slope, so one variable decreases as the other increases, the same should apply but the correlation should be negative. So we want to define a number that captures how closely the data are associated with each other, and in which direction.

An appropriate statistical measure was defined in its current form by the English mathematician and biostatistician Karl Pearson.[25] Pearson introduced the correlation coefficient. Given two random variables, find their means. Convert each to its difference from its mean, multiply them together, and calculate the expected value of this product. Finally, divide by the product of the standard deviations of the two random variables. The idea is that if the data are the same, the result is 1; if they are exact opposites, one being minus the other, the result is −1; and if they're independent, the result is 0. More generally, any *exact* linear relation leads either to 1 or −1, depending on the sign of the slope.

If two variables have a causal link – one causes the other – they should be highly correlated. When the data showed a large positive correlation between smoking cigarettes and developing lung cancer, doctors began to believe one caused the other. However, the same correlation arises whichever of the two variables is considered to be the potential cause. Perhaps a predisposition to lung cancer causes people to smoke more heavily. You can even invent a reason; for example, smoking might help to combat irritation of the lungs caused by precancerous cells. Or maybe something else causes both – perhaps stress. Whenever medical researchers discover large correlations between some product and a disease, the phrase 'correlation is not causality' is trotted out by companies that would lose profits if the public realised that their product does in fact cause the disease. It's true that correlation doesn't always imply causality, but the statement hides an inconvenient truth: correlation is a useful indicator of *potential* causality. Moreover, if there's independent evidence about how the product might lead to the disease, a high correlation can strengthen that evidence. When it was discovered that tobacco smoke contains carcinogens – cancer-inducing substances – the scientific case for causality became much stronger.

One way to sort out which of a number of potential influences are significant is to employ generalisations, such as the correlation matrix, which produces an array of correlation coefficients for many different data sets. Correlation matrices are useful for spotting connections, but they can be misused. For example, suppose you want to find out how diet affects a whole range of diseases. You draw up a list of 100 types of food and 40 diseases. Then you choose a sample of people, and find out which foods they eat and which diseases they've contracted. For each combination of food and disease you calculate the corresponding correlation coefficient: how closely associated that food and that disease are, across the sample of people. The resulting correlation matrix is a rectangular table, whose 100 rows correspond to foodstuffs and 40 columns to diseases. The entry in a given row and column is the correlation coefficient between the foodstuff corresponding to the row and the disease corresponding to the column. It has 4000 numbers in it. Now you look at the table and try to spot numbers close to 1, which indicate potential relationships between food and disease. Maybe the 'carrot' row and 'headache' column has an entry of 0·92, say. This leads you to infer, tentatively, that eating carrots *might* cause headaches.

What you ought to do now is to start an entirely new study, with new subjects, and collect new data to test that hypothesis. However, that costs money, so sometimes researchers extract the data from the original experiment. Then they use statistical tests to assess the significance of that particular correlation, ignoring all the other data and analysing just that one relationship, as if nothing else had been measured. Then they conclude that there's a significant chance that eating carrots causes headaches.

This approach is called 'double-dipping', and as presented it's fallacious. For example, suppose you select a woman at random, measure her height, and find that on the assumption of a normal distribution, the probability of anyone having that height (or more) is 1%. Then you'd be reasonably justified in concluding that her unusual height is not the result of random causes. However, if in fact you chose her by measuring the heights of hundreds of women and selecting the tallest, that conclusion wouldn't be justified: by pure chance, you're quite likely to get such a person in a population that big. Double-dipping with a correlation matrix is similar, but in a more complicated setting.

IN 1824 THE *ARU PENNSYLVANIAN* conducted a 'straw poll', an unofficial vote, to get some idea whether Andrew Jackson or John Quincy Adams would be elected president of the USA. The poll showed 335 votes for Jackson and 169 for Adams. Jackson won. Ever since, elections have attracted opinion pollsters. The polls (or 'surveys') sample only a small section of the population of voters, for practical reasons. So an important question arises: how big should the sample be to give accurate results? The same question is important in many other areas, such as census taking and medical trials of a new drug.

In Chapter 5 we saw that Laplace investigated sampling, but his recommendation for getting an accurate result was to make sure that data for the sample have similar proportions to data for the whole population. That can be hard to arrange, so pollsters (until recently) mainly focused on random samples, where people are chosen using some random process. Suppose, for example, that we want to find out the average size of a family, in a large population. We choose a random sample, and calculate the sample mean: the average size for the sample. Presumably, the bigger the sample, the closer the sample mean will be to the actual mean. How big a sample should we choose to be sufficiently certain that we've attained a given level of accuracy?

The mathematical setting associates a random variable with each family in the sample. It's assumed that the same probability distribution applies to each family – the distribution for the entire population – and we want to estimate its mean. The law of large numbers tells us that if the sample is large enough, the mean of the sample 'almost surely' gets as close as we wish to the true mean. That is, it does so with probability tending to 1 as the sample size increases without limit. But this doesn't tell us how big the sample should be. To find that, we need a more sophisticated result, the central limit theorem of Chapter 5, which relates the difference between the sample mean and the actual mean to a normal distribution.[26] We then use the normal distribution to calculate the relevant probabilities, and deduce the smallest sample size that ought to work.

In the example of family sizes, we first do a preliminary sample to estimate the standard deviation. A ballpark figure is enough. We decide how confident we want to be that our result is correct (say 99%) and how big an error we're willing to accept (say 1/10). It turns out that the sample size should be such that, assuming a standard normal distribution with mean 0 and standard deviation 1, the probability that the sample mean deviates from the true mean by less than 1/10 is at least 99%. Using the maths of the normal distribution, we find that the sample size should be at least $660\sigma^2$, where σ^2 is the variance for the whole population. Since we estimated that variance by a ballpark sample, we ought to allow for a bit more error, so we make the sample size a bit bigger than that. Notice that here the sample size doesn't depend on the size of the whole population. It depends on the variance of the random variable – how scattered it is.

For different sampling questions, similar analyses are employed, depending on the appropriate distribution and the quantity being estimated.

OPINION POLLS OCCUPY A SPECIAL area within sampling theory. The advent of social media has altered how many of them are carried out. Well-designed internet polls use a panel of carefully selected individuals, and ask for their views. However, many polls just let anyone who wants to vote do so. These polls are poorly designed, because people with strong views are more likely to vote, many people don't even know about the poll, and some may not have an internet connection, so the sample isn't representative. Telephone polls are also likely to be biased because many people don't answer cold-callers, or refuse to respond to a pollster when asked for their opinion. In this scam-ridden era, they may not even be sure that the call is a genuine poll. People may not have a phone. Some people don't tell the pollster their true intentions – for example, they may not be willing to tell a stranger that they plan to vote for an extremist party. Even the way a question is worded can affect how people respond.

Polling organisations use a variety of methods to try to minimise these sources of error. Many of these methods are mathematical, but psychological and other factors are also involved. There have been plenty of horror stories where polls have confidently indicated the wrong result, and it seems to be happening more often. Special factors are sometimes invoked to 'explain' why, such as a sudden late swing in opinion, or people deliberately lying to make the opposition think its ahead and get complacent. It's difficult to assess how valid these excuses are. Nevertheless, polling has a fairly good track record overall when performed competently, so it provides a useful tool for reducing uncertainty. Polls also runs the risk of influencing the outcome; for example, people may decide not to bother voting if they think their choice is bound to win. Exit polls, where people are asked who they voted for soon after they cast their vote, are often very accurate, giving the correct result long before the official vote count reveals it, and can't influence the result.

8

HOW CERTAIN ARE YOU?

Absurdity, *n*. A statement or belief manifestly inconsistent with one's own opinion.
Ambrose Bierce, *The Devil's Dictionary*

WHEN NEWTON PUBLISHED THE CALCULUS Bishop George Berkeley was moved to respond with a pamphlet: *The Analyst*, subtitled 'A discourse addressed to an infidel mathematician. Wherein it is examined whether the object, principles, and inferences of the modern analysis are more distinctly conceived, or more evidently deduced, than religious mysteries and points of faith.' The title page has a biblical epigram: 'First cast out the beam out of thine own Eye; and then shalt thou see clearly to cast out the mote out of thy brother's eye.—Matt. vii:5'

You don't need to be terribly sensitive to deduce that the bishop wasn't a calculus fan. When his pamphlet appeared in 1734, science was making big advances, and many scholars and philosophers were starting to argue that evidence-based science was superior to faith as a way to understand the natural world. The place of Christian beliefs, previously considered to be absolute truths by virtue of the authority of God, was being usurped by mathematics, which was not only true, but *necessarily* true, and it could prove it.

Of course mathematics is no such thing, but then neither is religion. However, at the time, the bishop had every reason to be sensitive about challenges to faith, and he set out to rectify matters by pointing out some logical difficulties in calculus. His not-very-hidden agenda was to convince the world that mathematicians weren't as logical as they claimed to be, demolishing their claim to be sole guardians of absolute truths. He had a point, but outright attack is a bad way to convince people they're wrong; mathematicians get upset when outsiders try to tell them how to do mathematics. Ultimately Berkeley had missed the point, and the experts could see that, even if at the time they couldn't quite lay down rigorous logical foundations.

THIS ISN'T A BOOK ABOUT calculus, but I'm telling the story because it leads directly to one of the great unappreciated heroes of mathematics, a man whose scientific reputation at the time of his death was very ordinary, but has gone from strength to strength ever since. His name was Thomas Bayes, and he created a revolution in statistics that has never been more relevant than it is today.

Bayes was born in 1701, perhaps in Hertfordshire. His father Joshua was a Presbyterian minister, and Thomas followed in his footsteps. He took a degree in logic and theology at Edinburgh University, assisted his father for a short period, and then became minister of the Mount Sion Chapel in Tunbridge Wells. He is the author of two very different books. The first, in 1731, was *Divine Benevolence, or an Attempt to Prove That the Principal End of the Divine Providence and Government is the Happiness of His Creatures.* Exactly what we might expect of a nonconformist man of the cloth. The other, in 1736, was *An Introduction to the Doctrine of Fluxions, and Defence of the Mathematicians against the Objections of the Author of the Analyst.* Which was most definitely not what we might expect. The minister Bayes was defending the scientist Newton against an attack by a bishop. The reason was simple: Bayes disagreed with Berkeley's mathematics.

When Bayes died, his friend Richard Price received some of his papers, and published two mathematical articles extracted from them. One was on asymptotic series – formulas approximating some important quantity by adding large numbers of simpler terms together, with a specific technical meaning for 'approximate'. The other bore the title 'Essay towards solving a problem in the doctrine of chances', and appeared in 1763. It was about conditional probability.

Bayes's key insight occurs early in the paper. Proposition 2 begins: 'If a person has an expectation depending on the happening of an event, the probability of the event is [in the ratio] to the probability of its failure as his loss if it fails [is in the ratio] to his gain if it happens.'

This is a bit of a mouthful, but Bayes explains it in more detail. I'll recast what he wrote in modern terms, but it's all there in his paper.

If E and F are events, write the conditional probability that E occurs, given that F has occurred, as $P(E|F)$ (read as 'the probability of E given F'). Assume the two events E and F are independent of each other. Then the formula that we now call Bayes's theorem states that

$$P(E|F) = \frac{P(F|E)P(E)}{P(F)}$$

This follows easily[27] from what is now considered to be the definition of conditional probability:

$$P(E|F) = \frac{P(E \text{ and } F)}{P(F)}$$

Let's check the second formula out on the two-girl puzzle, in the first case where we're told that at least one of the Smiths' children is a girl. The full sample space has four events GG, GB, BG, BB, each with probability 1/4. Let E be 'two girls', that is, GG. Let F be 'not two boys', which is the subset {GG, GB, BG}, probability 3/4. The event 'E and F' is GG, the same as event E. According to the formula, the probability that they have two girls, given that at least one child is a girl, is

$$P(E|F) = \frac{P(F|E)P(E)}{P(F)} = \frac{1/4}{3/4} = \frac{1}{3}$$

This is the result we got before.

Bayes went on to consider more complicated combinations of conditions and their associated conditional probabilities. These results, and more extensive modern generalisations, are also referred to as Bayes's theorem.

BAYES'S THEOREM HAS IMPORTANT PRACTICAL consequences, for example to quality control in manufacturing. There it provides answers to questions such as: 'Given that the wheels fell off one of our toy cars, what is the probability that it was manufactured at our Wormingham factory?' But over the years, it morphed into an entire philosophy about what probability *is,* and how to deal with it.

The classical definition of probability – 'interpretation' is perhaps a better word – is the frequentist one: the frequency with which an event occurs when an experiment is repeated many times. As we've seen, this interpretation goes right back to the early pioneers. But it has several defects. It's not clear exactly what 'many times' means. In some (rare) series of trials, the frequency need not converge to any well-defined number. But the main one is that it relies on being able to perform the identical experiment as many times as we wish. If that's not possible, it's unclear whether 'probability' has any meaning, and if it does, we don't know how to find it.

For example, what is the probability that we will discover intelligent aliens by the year 3000 AD? By definition, this is an experiment we can run only once. Yet most of us intuitively feel that this probability *ought* to have a meaning – even if we disagree entirely on its value. Some will insist the probability is 0, others 0·99999, and wishy-washy fence-sitters will go for 0·5 (the fifty-fifty default, almost certainly wrong). A few will trot out the Drake equation, whose variables are too imprecise to help much.[28]

The main alternative to frequentism is the Bayesian approach. Whether the good Reverend would recognise it as his brainchild is not totally clear, but he certainly set himself up for getting the (in some eyes dubious) credit. The approach actually goes back further, to Laplace, who discussed questions such as: 'How likely is it that the Sun will rise tomorrow?'. But we're stuck with the term now, if only for historical reasons.

Here's how Bayes defines probability in his paper: 'The probability of any event is the ratio between the value at which an expectation depending on the happening of the event ought to be computed, and the chance of the thing expected upon its happening.' This statement is somewhat ambiguous. What is our 'expectation'? What does 'ought to be' mean? One reasonable interpretation is that the probability of some event can be interpreted as our *degree of belief* that it will happen. How confident we are in it as a hypothesis; what odds would persuade us to gamble on it; how strongly we believe in it.

This interpretation has advantages. In particular, it lets us assign probabilities to events that can occur only once. My question about aliens could reasonably be answered 'The probability of intelligent aliens visiting us by the year 3018 is 0·316.' This statement doesn't mean 'If we run that period of history repeatedly a thousand times, aliens will turn up in 316 of the runs.' Even if we owned a time machine, and its use didn't change history, we'd get either 0 occasions of alien invasions, or 1000. No: the 0·316 means that our confidence that they'll turn up is moderate.

The interpretation of probability as degree of belief also has evident disadvantages. As George Boole wrote in *An Investigation into the Laws of Thought* in 1854, 'It would be unphilosophical to affirm that the strength of expectation, viewed as an emotion of the mind, is capable of being referred to any numerical standard. The man of sanguine temperament builds high hopes where the timid despair, and the irresolute are lost in doubt.' In other words, if you disagree with my assessment, and say that the probability of an alien invasion is only 0·003, there's no way to work out who's right. If anyone. That's true *even if the aliens turn up*. If they don't, your estimate is better than mine; if they do, mine is better than yours. But neither of us is demonstrably correct, and other figures would have performed better than either – depending on what happens.

Bayesians have a kind of answer to that objection. It reintroduces the possibility of repeating the experiment, though not under the exact selfsame conditions. Instead, we wait another thousand years, and see if another bunch of aliens turns up. But before doing that, we *revise* our degrees of belief.

Suppose, for the sake of argument, that an expedition from Apellobetnees III turns up in the year 2735. Then my 0·316 performed better than your 0·003. So, when it comes to the following millennium, we both revise our degrees of belief. Yours definitely needs revising upwards; maybe mine does too. Perhaps we agree to compromise on 0·718.

We can stop there. The exact event is not repeatable. We still have a better estimate of whatever it is we're estimating. If we're more ambitious, we can revise the question to 'probability of aliens arriving within a period of 1000 years', and do the experiment again. This time, oh dear, no new aliens arrive. So we revise our degree of belief down again, say to 0·584, and wait another thousand years.

That all sounds a bit arbitrary, and as stated it is. The Bayesian version is more systematic. The idea is that we start out with an initial degree of belief – the *prior* probability. We do the experiment (wait for aliens), observe the outcome, and use Bayes's theorem to calculate a *posterior* probability – our improved, better informed degree of belief. No longer just a guess, it's based on some limited evidence. Even if we stop there, we've achieved something useful. But where appropriate, we can reinterpret that posterior probability as a new prior one. Then we can do a second experiment to get a second posterior probability, which in some sense ought to be better still. Using that as a new prior, we experiment again, get an even better posterior ... and so it goes.

It still sounds subjective, which it is. Remarkably, however, it often works amazingly well. It gives results, and suggests methods, that aren't available in a frequentist model. Those results and methods can solve important problems. So the world of statistics has now fragmented into two distinct sects, Frequentists v. Bayesians, with two distinct ideologies: Frequentism and Bayesianism.

A pragmatic view would be that we don't have to choose. Two heads are better than one, two interpretations are better than one, and two philosophies are better than one. If one's no good, try the other. Gradually that view is prevailing, but right now a lot of people insist that only one sect is right. Scientists in many areas are entirely happy to go along with both, and Bayesian methods are widespread because they're more adaptable.

A COURT OF LAW MIGHT SEEM an unlikely test ground for mathematical theorems, but Bayes's theorem has important applications to criminal prosecutions. Unfortunately, the legal profession largely ignores this, and trials abound with fallacious statistical reasoning. It's ironic – but highly predictable – that in an area of human activity where the reduction of uncertainty is vital, and where well-developed mathematical tools exist to achieve just that, both prosecution and defence prefer to resort to reasoning that is archaic and fallacious. Worse, the legal system itself discourages the use of the mathematics. You might think that applications of probability theory in the courts should be no more controversial than using arithmetic to decide how much faster than the speed limit someone is driving. The main problem is that statistical inference is open to misinterpretation, creating loopholes that both prosecution and defence lawyers can exploit.

An especially devastating judgement against the use of Bayes's theorem in legal cases was delivered in the 1998 appeal of *Regina v. Adams*, a rape case where the sole evidence of guilt was a DNA match to a swab taken from the victim. The defendant had an alibi, and didn't resemble the victim's description of her attacker, but he was convicted because of this match. On appeal, the defence countered the prosecution's contention that the match probability was one in 200 million with testimony from an expert witness, who explained that any statistical argument must take account of the defence's evidence as well, and that Bayes's theorem was the correct approach. The appeal succeeded, but the judge condemned *all* statistical reasoning: 'The task of the jury is ... to evaluate evidence and reach a conclusion not by means of a formula, mathematical or otherwise, but by the joint application of their individual common sense and knowledge of the world to the evidence before them.' All very well, but Chapter 6 shows how useless 'common sense' can be in such circumstances.

Nulty & Ors v. Milton Keynes Borough Council in 2013 was a civil case about a fire at a recycling centre near Milton Keynes. The judge concluded that the cause was a discarded cigarette, because the alternative explanation – electrical arcing – was even less likely. The company insuring the engineer who allegedly threw away the cigarette lost the case and was told to pay £2 million compensation. The Appeal Court rejected the judge's reasoning, but disallowed the appeal. The judgement threw out the entire basis of Bayesian statistics: 'Sometimes the "balance of probability" standard is expressed mathematically as "50+ % probability", but this can carry with it a danger of pseudo-mathematics... To express the probability of some event having happened in percentage terms is illusory.'

Norma Fenton and Martin Neil[29] report a lawyer saying: 'Look, the guy either did it or he didn't do it. If he did then he is 100% guilty and if he didn't then he is 0% guilty; so giving the chances of guilt as a probability somewhere in between makes no sense and has no place in the law.' It's unreasonable to assign probabilities to events you *know* have (or have not) happened. But it's entirely sensible to assign probabilities when you don't know, and Bayesianism is about how to do this rationally. For instance, suppose someone tosses a coin; they look at it, you don't. To them, the outcome is known and its probability is 1. But to you, the probability is 1/2 for each of heads and tails, because you're not assessing what happened, you're assessing how likely your guess is to be right. In *every* court case, the defendant is either guilty or not – but that information has no bearing on the court, whose job to find out *which*. Rejecting a useful tool because it might bamboozle juries is a bit silly if you allow them to be bamboozled in the time-honoured manner by lawyers talking rubbish.

MATHEMATICAL NOTATION CONFUSES MANY PEOPLE, and it certainly doesn't help that our intuition for conditional probabilities is so poor. But those aren't good reasons to reject a valuable statistical tool. Judges and juries routinely deal with highly complex circumstances. The traditional safeguards are things like expert witnesses – though as we'll see, their advice is not infallible – and careful direction of the jury by the judge. In the two cases I mentioned, it was, perhaps, reasonable to rule that the lawyers hadn't presented a sufficiently compelling statistical case. But ruling against the future use of anything remotely related was, in the eyes of many commentators, a step too far, making it much harder to convict the guilty, thereby reducing the protection of the innocent. So a brilliant and fundamentally simple discovery that had proved to be a very useful tool for reducing uncertainty went begging, because the legal profession either didn't understand it, or was willing to abuse it.

Unfortunately, it's all too easy to abuse probabilistic reasoning, especially about conditional probabilities. We saw in Chapter 6 how easily our intuition can be led astray, and that's when the mathematics is clear and precise. Imagine you're in court, accused of murder. A speck of blood on the victim's clothing has been DNA-fingerprinted, and there's a close match to your DNA. The prosecution argues that the match is so close that the chances of it occurring for a randomly chosen person are one in a million – which may well be true, let's assume it is – and concludes that the chance you are innocent is also one in a million. This, in its simplest form, is the prosecutor's fallacy. It's total nonsense.

Your defence counsel springs into action. There are sixty million people in the United Kingdom. Even with those one-in-a-million odds,

60 of them are equally likely to be guilty. Your chance of guilt is 1/60, or about 1·6%. This is the defence lawyer's fallacy, and it's also nonsense.

These examples are invented, but there are plenty of cases in which something along those lines has been put to the court, among them *Regina v. Adams*, where the prosecution highlighted an irrelevant DNA match probability. There are clear examples where innocent people have been convicted on the basis of the prosecutor's fallacy, as the courts themselves have acknowledged by reversing the conviction on appeal. Experts in statistics believe that many comparable miscarriages of justice, caused by fallacious statistical reasoning, have not been reversed. It seems likely, though harder to prove, that criminals have been found not guilty because the court fell for the defence lawyer's fallacy. It's easy enough, however, to explain why both lines of reasoning are fallacious.

For present purposes, let's leave aside the issue of whether probabilistic calculations should be allowed in trials at all. Trials are, after all, supposed to assess guilt or innocence, not to convict you on the grounds that you *probably* did the dirty deed. The topic under discussion is what we need to be careful about when the use of statistics *is* allowed. As it happens, there's nothing to stop probabilities being presented as evidence in British or American law. Clearly the prosecution and the defence in my DNA scenario can't both be right, since their assessments disagree wildly. So what's wrong?

The plot of Conan Doyle's short story 'Silver Blaze' hinges on Sherlock Holmes drawing attention to 'the curious incident of the dog in the night-time'. 'The dog did *nothing* in the night-time,' protests Scotland Yard's Inspector Gregory. To which Holmes, enigmatic as ever, replies, 'That was the curious incident.' There's a dog doing nothing in the two arguments above. What is it? There's no reference to any other evidence that might indicate guilt or innocence. But additional evidence has a strong effect on the *a priori* probability that you're guilty, and it changes the calculations.

Here's another scenario that may help to make the problem clear. You get a phone call telling you you've won the National Lottery, a cool 10 million pounds. It's genuine, and you pick up your winner's cheque. But when you present it to your bank, you feel a heavy hand on your shoulder: it's a policeman, and he arrests you for theft. In court, the prosecution argues that you almost certainly cheated, defrauding the lottery company of the prize money. The reason is simple: the chance of any randomly chosen person winning the lottery is one in 20 million. By the prosecutor's fallacy, that's the probability that you're innocent.

In this case it's obvious what's wrong. Tens of millions of people play the lottery every week; *someone* is very likely to win. You haven't been selected at random beforehand; you've been selected after the event *because you won*.

AN ESPECIALLY DISTURBING LEGAL CASE involving statistical evidence occurred in the trial of Sally Clark, a British lawyer who lost two children to cot deaths (SIDS, sudden infant death syndrome). The prosecution's expert witness testified that the probability of this double tragedy occurring by accident was one in 73 million. He also said that the actual rate observed was more frequent, and he explained this discrepancy by offering the opinion that many double cot deaths weren't accidental, but resulted from Munchausen syndrome by proxy, his own special area of expertise. Despite the absence of any significant corroborating evidence other than the statistics, Clark was convicted of murdering her children, widely reviled in the media, and sentenced to life imprisonment.

Serious flaws in the prosecution's case were evident from the outset, to the alarm of the Royal Statistical Society, which pointed them out in a press release after the conviction. After more than three years in prison Clark was released on appeal, but not because of any of those flaws: because the pathologist who had examined the babies after the deaths had withheld possible evidence of her innocence. Clark never recovered from this miscarriage of justice. She developed psychiatric problems and died of alcohol poisoning four years later.

The flaws are of several kinds. There's clear evidence that SIDS has a genetic element, so one cot death in a family makes a second one more likely. You can't sensibly estimate the chances of two successive deaths by multiplying the chance of one such death by itself. The two events are *not independent*. The claim that most double deaths are the result of Munchausen syndrome by proxy is open to challenge. Munchausen syndrome is a form of self-harm. Munchausen syndrome *by proxy* is self-harm inflicted by harming someone else. (Whether this makes sense is controversial.) The court seemed unaware that the higher rate of double deaths reported by their expert witness might, in fact, just be the *actual* rate of accidental ones. With, no doubt, a few rare cases where the children really had been murdered.

But all this is beside the point. Whatever probability is used for accidental double cot deaths, it has to be compared with the possible alternatives. Moreover, everything has to be conditional on the other evidence: *two deaths have occurred.* So there are three possible explanations: both deaths were accidents; both were murders; something else entirely (such as one murder, one natural death). All three events are extremely unlikely: if anything probabilistic matters, it's how unlikely each is compared with the others. And even if either death was a murder, there also remains the question: *Who did it?* It's not automatically the mother.

So the court focused on:

■ The probability that a randomly chosen family suffers two cot deaths

when it should have been thinking about:

■ The probability that the mother is a double murderer, given that two cot deaths have occurred.

It then confused the two, while using incorrect figures.

Ray Hill, a mathematician, carried out a statistical analysis of cot deaths, using real data, and found that it's between 4·5 and 9 times as likely that there will be two SIDS accidents in one family as it is that there will be two murders. In other words, on statistical grounds alone, the likelihood of Clark being guilty was only 10–20%.

Again, the dog didn't bark. In this kind of case, statistical evidence alone is totally unreliable unless supported by other lines of evidence. For example, it would have improved the prosecution's case if it could have been established, independently, that the accused had a track record of abusing her children, but no such track record existed. And it would have improved the defence's case to argue that there had been no signs of such abuse. Ultimately, the sole 'evidence' of her guilt was that two children had died, apparently from SIDS.

Fenton and Neil discuss a large number of other cases in which statistical reasoning may have been misapplied.[30] In 2003 Lucia de Berk, a Dutch children's nurse, was accused of four murders and three attempted murders. An unusually high number of patients had died while she was on duty in the same hospital, and the prosecution put together circumstantial evidence, claiming that the probability of this happening by chance was one in 342 million. This calculation referred to the probability of the evidence existing, given that the accused is not guilty. What should have been calculated was the probability of guilt, given the evidence. De Berk was convicted and sentenced to life imprisonment. On appeal, the verdict was upheld, even though alleged evidence of her guilt was retracted during the appeal by a witness who admitted: 'I made it up.' (This witness was being detained in a criminal psychology unit at the time.) The press, not surprisingly, disputed the conviction, a public petition was organised, and in 2006 the Netherlands Supreme Court sent the case back to the Amsterdam Court, which again upheld the conviction. In 2008, after much bad publicity, the Supreme Court reopened the case. In 2010 a retrial found that all of the deaths had been due to natural causes and that the nurses involved had saved several lives. The court quashed the conviction.

It ought to be obvious that with large numbers of deaths in hospitals, and large numbers of nurses, unusually strong associations between some deaths and a particular nurse are likely. Ronald Meester and colleagues[31] suggest that the 'one in 342 million' figure is a case of double-dipping (see Chapter 7). They show that more appropriate statistical methods lead to figure of about one in 300, or even one in 50. These values are not statistically significant as evidence of guilt.

IN 2016 FENTON, NEIL, AND Daniel Berger published a review of Bayesian reasoning in legal cases. They analysed why the legal profession is suspicious of such arguments and reviewed their potential. They began by pointing out that the use of statistics in legal proceedings has increased considerably over the past four decades, but that most such uses involve classical statistics, even though the Bayesian approach avoids many of the pitfalls associated with classical methods, and is more widely applicable. Their main conclusions were that the lack of impact stems from 'misconceptions by the legal community about Bayes's theorem ... and the lack of adoption of modern computational methods'. And they advocated the use of a new technique, Bayesian networks, which could automate the calculations required in a way that would 'address most concerns about using Bayes in the law'.

Classical statistics, with its rather rigid assumptions and long-standing traditions, is open to misinterpretation. An emphasis on statistical significance tests can lead to the prosecutor's fallacy, because the probability of the evidence given guilt can be misrepresented as the probability of guilt given the evidence. More technical concepts such as confidence intervals, which define the range of values in which we can confidently assume some number lies, are 'almost invariably misinterpreted since their proper definition is both complex and counterintuitive (indeed it is not properly understood even by many trained statisticians).' These difficulties, and the poor track record of classical statistics, made lawyers unhappy about all forms of statistical reasoning.

This may be one reason for resisting Bayesian methods. Fenton and colleagues suggest another, more interesting one: too many of the Bayesian models presented in court are oversimplified. These models are used because of the assumption that the calculations involved should be sufficiently simple that they can be done by hand, so the judge and jury can follow them.

In the computer age, this restriction isn't necessary. It's sensible to be concerned about incomprehensible computer algorithms; as an extreme case we can imagine an artificially intelligent Justice Computer that silently weights the evidence and declares either 'guilty' or 'not guilty', without explanation. But when the algorithm is entirely comprehensible, and the sums are straightforward, it ought not to be hard to safeguard against the more obvious potential issues.

The toy models of Bayesian reasoning that I've discussed involve a very small number of statements, and all we've done is think about how probable one statement is, given some other statement. But a legal case involves all sorts of evidence, along with the statements associated with them, such as 'the suspect was at the crime scene', 'the subject's DNA matches the traces of blood on the victim', or 'a silver car was seen in the vicinity'. A Bayesian network represents all of these factors, and how they influence each other, as a directed graph: a collection of boxes joined by arrows. There's one box for each factor and one arrow for each influence. Moreover, associated with each arrow is a number, the conditional probability of the factor at its head, given the factor at its tail. A generalisation of Bayes's theorem then makes it possible to compute the probability of any particular factor occurring, given any other, or even given everything else that's known.

Fenton and colleagues suggest that Bayesian networks, suitably implemented, developed, and tested, could become an important legal tool, able to 'model the correct relevant hypotheses and the full causal context of the evidence'. Certainly there are plenty of issues about what types of evidence can be deemed suitable for such treatment, and those issues need to be debated and agreed. That said, the main obstacles preventing such a debate are the strong cultural barriers between science and the law that currently exist.

9

LAW AND DISORDER

Heat won't pass from a cooler to a hotter,
You can try it if you like but you far better notter.
Michael Flanders and Donald Swann, *First and Second Law* (lyrics)

FROM LAW AND ORDER TO law and disorder. From human affairs to physics.

One of the very few scientific principles to have become a household name, or at least to have come close, is the second law of thermodynamics. In his notorious 'two cultures' Rede lectures of 1959, and the subsequent book, the novelist C.P. Snow said that no one should consider themselves cultured if they don't know what this law says:

> A good many times I have been present at gatherings of people who, by the standards of the traditional culture, are thought highly educated and who have with considerable gusto been expressing their incredulity at the illiteracy of scientists. Once or twice I have been provoked and have asked the company how many of them could describe the Second Law of Thermodynamics. The response was cold: it was also negative. Yet I was asking something which is the scientific equivalent of: *Have you read a work of Shakespeare's?*

He was making a sensible point: basic science is at least as much a part of human culture as, say, knowing Latin quotations from Horace, or being able to quote verse by Byron or Coleridge. On the other hand, he really ought to have chosen a better example, because plenty of *scientists* don't have the second law of thermodynamics at their fingertips.[32]

To be fair, Snow went on to suggest that no more than one in ten educated people would be able to explain the meaning of simpler concepts such as mass or acceleration, the scientific equivalent of asking 'Can you read?' Literary critic F.R. (Frank Raymond) Leavis replied that there was only one culture, *his* – inadvertently making Snow's point for him.

In a more positive response, Michael Flanders and Donald Swann wrote one of the much-loved comic songs that they performed in their touring revues *At the Drop of a Hat* and *At the Drop of Another Hat* between 1956 and 1967, two lines of which appear as an epigram at the start of this chapter.[33] The scientific statement of the second law is phrased in terms of a rather slippery concept, which occurs towards the end of the Flanders & Swann song: 'Yeah, that's entropy, man.'

Thermodynamics is the science of heat, and how it can be transferred from one object or system to another. Examples are boiling a kettle, or holding a balloon over a candle. The most familiar thermodynamic variables are temperature, pressure, and volume. The ideal gas law tells us how they're related: pressure times volume is proportional to absolute temperature. So, for instance, if we heat the air in a balloon, the temperature increases, so either it has to occupy a greater volume (the balloon expands), or the pressure inside has to increase (eventually bursting the balloon), or a bit of both. Here I'm ignoring the obvious point that the heat may also burn or melt the balloon, which is outside the scope of the ideal gas law.

Another thermodynamic variable is heat, which is distinct from, and in many respects simpler than, temperature. Far subtler than either is entropy, which is often described informally as a measure of how disordered a thermodynamic system is. According to the second law, in any system not affected by external factors, entropy always increases. In this context, 'disorder' is not a definition but a metaphor, and one that's easily misinterpreted.

The second law of thermodynamics has important implications for the scientific understanding of the world around us. Some are cosmic in scale: the heat death of the universe, in which aeons into the future everything has become a uniform lukewarm soup. Some are misunderstandings, such as the claim that the second law makes evolution impossible because more complex organisms are more ordered. And some are very puzzling and paradoxical: the 'arrow of time', in which entropy seems to single out a specific forward direction for the flow of time, even though the equations from which the second law is derived are the same, in whichever direction time flows.

The theoretical basis for the second law is kinetic theory, introduced by the Austrian physicist Ludwig Boltzmann in the 1870s. This is a simple mathematical model of the motion of molecules in a gas, which are represented as tiny hard spheres that bounce off each other if they collide. The molecules are assumed to be very far apart, on average, compared to their size – not packed tightly as in a liquid or, even more so, a solid. At the time, most leading physicists didn't believe in molecules. In fact, they didn't believe that matter is made of atoms, which combine to make molecules, so they gave Boltzmann a hard time. Scepticism towards his ideas continued throughout his career, and in 1906 he hanged himself while on holiday. Whether resistance to his ideas was the cause, it's difficult to say, but it was certainly misguided.

A central feature of kinetic theory is that in practice the motion of the molecules looks random. This is why a chapter on the second law of thermodynamics appears in a book on uncertainty. However, the bouncing spheres model is deterministic, and the motion is chaotic. But it took over a century for mathematicians to prove that.[34]

THE HISTORY OF THERMODYNAMICS AND kinetic theory is complicated, so I'll omit points of fine detail and limit the discussion to gases, where the issues are simpler. This area of physics went through two major phases. In the first, classical thermodynamics, the important features of a gas were macroscopic variables describing its overall state: the aforementioned temperature, pressure, volume, and so on. Scientists were aware that a gas is composed of molecules (although this remained controversial until the early 1900s), but the precise positions and speeds of these molecules were not considered as long as the overall state was unaffected. Heat, for example, is the total kinetic energy of the molecules. If collisions cause some to speed up, but others to slow down, the total energy stays the same, so such changes have no effect on these macroscopic variables. The mathematical issue was to describe how the macroscopic variables are related to each other, and to use the resulting equations ('laws') to deduce how the gas behaves. Initially, the main practical application was to the design of steam engines and similar industrial machinery. In fact, the analysis of theoretical limits to the efficiency of steam engines motivated the concept of entropy.

In the second phase, microscopic variables such as the position and velocity of the individual molecules of the gas took precedence. The first main theoretical problem was to describe how these variables change as the molecules bounce around inside the container; the second was to deduce classical thermodynamics from this more detailed microscopic picture. Later, quantum effects were taken into consideration as well with the advent of quantum thermodynamics, which incorporates new concepts such as 'information' and provides detailed underpinnings for the classical version of the theory.

In the classical approach, the entropy of a system was defined indirectly. First, we define how this variable *changes* when the system itself changes; then we add up all those tiny changes to get the entropy itself. If the system undergoes a small change of state, the change in entropy is the change in heat divided by the temperature. (If the state change is sufficiently small, the temperature can be considered to be constant during the change.) A big change of state can be considered as a large number of small changes in succession, and the corresponding change in the entropy is the sum of all the small changes for each step. More rigorously, it's the integral of those changes in the sense of calculus.

That tells us the change in entropy, but what about the entropy itself? Mathematically, the change in entropy doesn't define entropy uniquely: it defines it except for an added constant. We can fix this constant by making a specific choice for the entropy at some well-defined state. The standard choice is based on the idea of absolute temperature. Most of the familiar scales for measuring temperature, such as degrees Celsius (commonly used in Europe) or Fahrenheit (America), involve arbitrary choices. For Celsius, 0°C is defined to be the melting point of ice, and 100°C is the boiling point of water. On the Fahrenheit scale, the corresponding temperatures are 32°F and 212°F. Originally Daniel Fahrenheit used the temperature of the human body for 100°F, and the coldest thing he could get as 0°F. Such definitions are hostages to fortune, whence the 32 and 212. In principle, you could assign any numbers you like to these two temperatures, or use something totally different, such as the boiling points of nitrogen and lead.

As scientists tried to create lower and lower temperatures, they discovered that there's a definite limit to how cold matter can become. It's roughly −273°C, a temperature called 'absolute zero' at which all thermal motion ceases in the classical description of thermodynamics. However hard you try, you can't make anything colder than that. The Kelvin temperature scale, named after the Irish-born Scots physicist Lord Kelvin, is a thermodynamic temperature scale that uses absolute zero as its zero point; its unit of temperature is the kelvin (symbol: K). The scale is just like Celsius, except that 273 is added to every temperature. Now ice melts at 273K, water boils at 373K, and absolute zero is 0K. The entropy of a system is now defined (up to choice of units) by choosing the arbitrary added constant so that the entropy is zero when the absolute temperature is zero.

That's the classical definition of entropy. The modern statistical mechanics definition is in some respects simpler. It boils down to the same thing for gases, though that's not immediately obvious, so there's no harm in using the same word in both cases. The modern definition works in terms of the microscopic states: microstates for short. The recipe is simple: if the system can exist in any of N microstates, all equally likely, then the entropy S is

$$S = k_B \log N$$

where k_B is a constant called Boltzmann's constant. Numerically, it's $1{\cdot}38065 \times 10^{-23}$ joules per kelvin. Here log is the natural logarithm, the logarithm to base $e = 2{\cdot}71828...$. In other words, the entropy of the system is proportional to the logarithm of the number of microstates that in principle it could occupy.

For illustrative purposes, suppose that the system is a pack of cards, and a microstate is any of the orders into which the pack can be shuffled. From Chapter 4 the number of microstates is 52!, a rather large number starting with 80,658 and having 68 digits. The entropy can be calculated by taking the logarithm and multiplying by Boltzmann's constant, which gives

$$S = 2{\cdot}15879 \times 10^{-21}$$

If we now take a second pack, it has the same entropy S. But if we combine the two packs, and shuffle the resulting larger pack, the number of microstates is now $N = 104!$, which is a much larger number starting 10,299 with 167 digits. The entropy of the combined system is now

$$T = 5{\cdot}27765 \times 10^{-21}$$

The sum of the entropies of the two subsystems (packs) before they were combined is

$$2S = 4{\cdot}31758 \times 10^{-21}$$

Since T is bigger than $2S$, the entropy of the combined system is larger than the sum of the entropies of the two subsystems.

Metaphorically, the combined pack represents all possible interactions between the cards in the two separate packs. Not only can we shuffle each pack separately; we can get extra arrangements by mixing them together. So the entropy of the system when interactions are permitted is bigger than the sum of the entropies of the two systems when they don't interact. The probabilist Mark Kac used to describe this effect in terms of two cats, each with a number of fleas. When the cats are separate, the fleas can move around, but only on 'their' cat. If the cats meet up, they can exchange fleas, and the number of possible arrangements increases.

Because the logarithm of a product is the sum of the logarithms of the factors, entropy increases whenever the number of microstates for the combined system is bigger than the product of the numbers of microstates of the individual systems. This is usually the case, because the product is the number of microstates of the combined system when the two subsystems are not allowed to mix together. Mixing allows more microstates.

Now imagine a box with a partition, a lot of oxygen molecules on one side, and a vacuum on the other. Each of these two separate subsystems has a particular entropy. Microstates can be thought of as the number of ways to arrange the positions of the separate molecules, provided we 'coarse-grain' space into a large but finite number of very tiny boxes, and use them to say where the molecules are. When the partition is removed, all the old microstates of the molecules still

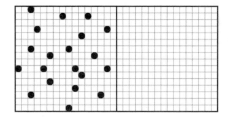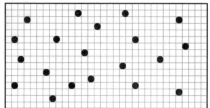

Left: With partition. *Right*: When the partition is removed, many new microstates are possible. Coarse-grained boxes in grey.

exist. But there are lots of new microstates, because molecules can go into the other half of the box. The new ways greatly outnumber the old ways, and with overwhelming probability the gas ends up filling the whole box with uniform density.

More briefly: the number of available microstates increases when the partition is removed, so the entropy – the logarithm of that number – also increases.

Physicists say that the state when the partition is still present is ordered, in the sense that the set of oxygen molecules in one part is kept separate from the vacuum in the other. When we remove the partition, this separation ceases, so the state becomes more disordered. This is the sense in which entropy can be interpreted as the amount of disorder. It's not a terribly helpful metaphor.

NOW WE COME TO THE vexed issue of the arrow of time.

The detailed mathematical model for a gas is a finite number of very tiny hard spherical balls, bouncing around inside a box. Each ball represents one molecule. It's assumed that collisions between molecules are perfectly elastic, which means that no energy is lost or gained in the collision. It's also assumed that when a ball collides with the walls it bounces like an idealised billiard ball hitting a cushion: it comes off at the same angle that it hit (no spin), but measured in the opposite direction, and it moves with exactly the same speed that it had before it hit the wall (perfectly elastic cushion). Again, energy is conserved.

The behaviour of those little balls is governed by Newton's laws of motion. The important one here is the second law: the force acting on a body equals its acceleration times its mass. (The first law says the body moves at uniform speed in a straight line unless a force acts on it; the third says that to every action there is an equal and opposite reaction.) When thinking about mechanical systems, we usually know the forces and want to find out how the particle moves. The law implies that at any given instant the acceleration equals the force divided by the mass. It applies to every one of those little balls, so in principle we can find out how all of them move.

The equation that arises when we apply Newton's laws of motion takes the form of a differential equation: it tells us the rate at which certain quantities change as time passes. Usually we want to know the quantities themselves, rather than how rapidly they're changing, but we can work out the quantities from their rates of change using integral calculus. Acceleration is the rate of change of velocity, and velocity is the rate of change of position. To find out where the balls are at any given instant, we find all the accelerations using Newton's law, then apply calculus to get their velocities, and apply calculus again to get their positions.

Two further ingredients are needed. The first is initial conditions. These specify where all the balls are at a given moment (say time $t = 0$) and how fast they're moving (and in which direction). This information pins down a unique solution of the equations, telling us what happens to that initial arrangement as time passes. For colliding balls, it all boils down to geometry. Each ball moves at a constant speed along a straight line (the direction of its initial velocity) until it hits another ball. The second ingredient is a rule for what happens then: they bounce off each other, acquiring new speeds and directions, and continue again in a straight line until the next collision, and so on. These rules determine the kinetic theory of gases, and the gas laws and suchlike can be deduced from them.

Newton's laws of motion for any system of moving bodies lead to equations that are time-reversible. If we take any solution of the equations, and run time backwards (change the time variable t to its negative $-t$) we also get a solution of the equations. It's usually not the *same* solution, though sometimes it can be. Intuitively, if you make a movie of a solution and run it backwards, the result is also a solution. For example, suppose you throw a ball vertically into the air. It starts quite fast, slows down as gravity pulls it, becomes stationary for an instant, and then falls, speeding up until you catch it again. Run the movie backwards and the same description applies. Or strike a pool ball with the cue so that it hits the cushion and bounces off; reverse the movie and again you see a ball hitting the cushion and bouncing off. As this example shows, allowing spheres to bounce doesn't affect reversibility, provided the bouncing rule works the same way when it's run backwards.

This is all very reasonable, but we've all seen reverse-time movies where weird things happen. Egg white and a yolk in a bowl suddenly rise into the air, are caught between two halves of a broken eggshell, which join together to leave an intact egg in the cook's hand. Shards of glass on the floor mysteriously move towards each other and assemble into an intact bottle, which jumps into the air. A waterfall flows *up* the cliff instead of falling down it. Wine rises up from a glass, back into the bottle. If it's champagne, the bubbles shrink and go back along with the wine; the cork appears mysteriously from some distance away and slams itself back into the neck of the bottle, trapping the wine inside. Reverse movies of people eating slices of cake are distinctly revolting – you can figure out what they look like.

Most processes in real life don't make much sense if you reverse time. A few do, but they're exceptional. Time, it seems, flows in only one direction: the arrow of time points from the past towards the future.

Of itself, that's not a puzzle. Running a movie backwards doesn't actually run *time* backwards. It just gives us an impression of what would happen if we could do that. But thermodynamics reinforces the irreversibility of time's arrow. The second law says that entropy increases as time passes. Run that backwards and you get something where entropy decreases, breaking the second law. Again, it makes sense; you can even define the arrow of time to be the direction in which entropy increases.

Things start to get tricky when you think about how this meshes with kinetic theory. The second law of Newton says the system is time-reversible; the second law of thermodynamics says it's not. But the laws of thermodynamics are a *consequence* of Newton's laws. Clearly something is screwy; as Shakespeare put it, 'the time is out of joint'.

The literature on this paradox is enormous, and much of it is highly erudite. Boltzmann worried about it when he first thought about the kinetic theory. Part of the answer is that the laws of thermodynamics are statistical. They don't apply to every single solution of Newton's equations for a million bouncing balls. In principle, all the oxygen molecules in a box could move into one half. Get that partition in, fast! But mostly, they don't; the probability that they do would be something like 0·000000..., with so many zeros before you reach the first nonzero digit that the planet is too small to hold them.

However, that's not the end of the story. For every solution such that entropy increases as time passes, there's a corresponding time-reversed solution in which entropy decreases as time passes. Very rarely, the reversed solution is the same as the original (the thrown ball, provided initial conditions are taken when it reaches to peak of its trajectory; the billiard ball, provided initial conditions are taken when it hits the cushion). Ignoring these exceptions, solutions come in pairs: one with entropy increasing; the other with entropy decreasing. It makes no sense to maintain that statistical effects select only one half. It's like claiming that a fair coin always lands heads.

Another partial resolution involves symmetry-breaking. I was very careful to say that when you time-reverse a solution of Newton's laws, you always get a solution, but not necessarily the same one. The time-reversal symmetry of the laws does not imply time-reversal symmetry of any given solution. That's true, but not terribly helpful, because solutions still come in pairs, and the same issue arises.

So why does time's arrow point only one way? I have a feeling that the answer lies in something that tends to get ignored. Everyone focuses on the time-reversal symmetry of the *laws*. I think we should consider the time-reversal asymmetry of the *initial conditions*.

That phrase alone is a warning. When time is reversed, initial conditions aren't initial. They're final. If we specify what happens at time zero, and deduce the motion for positive time, we've already fixed a direction for the arrow. That may sound a silly thing to say when the mathematics also lets us deduce what happens for negative time, but hear me out. Let's compare the bottle being dropped and smashed with the time-reversal: the broken fragments of bottle unsmash and reassemble.

In the 'smash' scenario, the initial conditions are simple: an intact bottle, held in the air. Then you let go. As time passes, the bottle falls, smashes, and thousands of tiny pieces scatter. The final conditions are very complicated, the ordered bottle has turned into a disordered mess on the floor, entropy has increased, and the second law is obeyed.

The 'unsmash' scenario is rather different. The initial conditions are complicated: lots of tiny shards of glass. They may seem to be stationary, but actually they're all moving very slowly. (Remember, we're ignoring friction.) As time passes, the shards move together and unsmash, to form an intact bottle that leaps into the air. The final conditions are very simple, the disordered mess on the floor has turned into the ordered bottle, entropy has decreased, and the second law is disobeyed.

The difference here has nothing to do with Newton's law, or its reversibility. Neither is it governed by entropy. Both scenarios are consistent with Newton's law; the difference comes from the choice of initial conditions. The 'smash' scenario is easy to produce experimentally, because the initial condition is easy to arrange: get a bottle, hold it up, let go. The 'unsmash' scenario is impossible to produce experimentally, because the initial condition is too complicated, and too delicate, to set up. In principle it exists, because we can solve the equations for the falling bottle until some moment after it has smashed. Then we take that state as initial condition, but with all velocities reversed, and the symmetry of the mathematics means that the bottle would indeed unsmash – but only if we realised those impossibly complicated 'initial' conditions exactly.

Our ability to solve for negative time also starts with the intact bottle. It calculates how it got to that state. Most likely, someone put it there. But if you run Newton's laws backwards in time, you don't deduce the mysterious appearance of a hand. The particles that made up the hand are absent from the model you're solving. What you get is a hypothetical past that is mathematically consistent with the chosen 'initial' state. In fact, since tossing a bottle into the air is its own time-reversal, the backwards solution would also involve the bottle falling to the ground, and 'by symmetry' it would also smash. But in reverse time. The full story of the bottle – not what actually happened, because the Hand of God at time zero isn't part of the model – consists of thousands of shards of glass, which start to converge, unsmash, rise in the air as an intact bottle, reach the peak of their trajectory at time zero, then fall, smash, and scatter into thousands of shards. Initially, entropy decreases; then it increases again.

In *The Order of Time*, Carlo Rovelli says something very similar.[35] The entropy of a system is defined by agreeing not to distinguish between certain configurations (what I called coarse-graining). So entropy depends on what information about the system we can access. In his view, we don't experience an arrow of time because that's the direction along which entropy increases. Instead, we think entropy increases because *to us* the past seems to have lower entropy than the present.

I SAID THAT SETTING UP INITIAL conditions to smash a bottle is easy, but there's a sense in which that's untrue. It's true if I can go down to the supermarket, buy a bottle of wine, drink it, and then use the empty bottle. But where did the bottle come from? If we trace its history, its constituent molecules probably went through many cycles of being recycled and melted down; they came from many different bottles, often smashed before or during recycling. But all of the glass involved must eventually trace back to sand grains, which were melted to form glass. The actual 'initial conditions' decades or centuries ago were at least as complicated as the ones I declared to be impossible for unsmashing the bottle.

Yet, miraculously, the bottle was made.

Does this disprove the second law of thermodynamics?

Not at all. The kinetic theory of gases – indeed, the whole of thermodynamics – involves simplifying assumptions. It models certain common scenarios, and the models are good when those scenarios apply.

One such assumption is that the system is 'closed'. This is usually stated as 'no external energy input is allowed', but really what we need is 'no external influence not built into the model is allowed'. The manufacture of a bottle from sand involves huge numbers of influences that are not accounted for if you track only the molecules in the bottle.

The traditional scenarios in the thermodynamics texts all involve this simplification. The text will discuss a box with a partition, containing gas molecules in one half (or some variant of this set-up). It will explain that if you *then* remove the partition, entropy increases.

But what it doesn't discuss is how the gas got into the box in that *initial* arrangement. Which had lower entropy than the gas did when it was part of the Earth's atmosphere. Agreed, that's no longer a closed system. But what really matters here is the type of system, or, more specifically, the initial conditions assumed. The mathematics doesn't tell us how those conditions were actually realised. Running the smashing-bottle model backwards doesn't lead to sand. So the model really only applies to forward time. I italicised two words in the above description: *then* and *initial*. In backward time, these have to change to *before* and *final*. The reason that time has a unique arrow in thermodynamics, despite the equations being time-reversible, is that the scenarios envisaged have an arrow of time built in: the use of *initial* conditions.

It's an old story, repeated throughout human history. Everyone is so focused on the *content* that they ignore the *context*. Here, the content is reversible, but the context isn't. That's why thermodynamics doesn't contradict Newton. But there's another message. Discussing a concept as subtle as entropy using vague words like 'disorder' is likely to lead to confusion.

10

UNPREDICTING THE PREDICTABLE

We demand rigidly defined areas of doubt and uncertainty!
Douglas Adams, *The Hitchhiker's Guide to the Galaxy*

IN THE 16TH AND 17TH centuries, two great figures of science noticed mathematical patterns in the natural world. Galileo Galilei observed them down on the ground, in the motion of rolling balls and falling bodies. Johannes Kepler discovered them in the heavens, in the orbital motion of the planet Mars. In 1687, building on their work, Newton's *Principia* changed how we think about nature by uncovering deep mathematical laws that govern nature's uncertainties. Almost overnight, many phenomena, from the tides to planets and comets, became predictable. European mathematicians quickly recast Newton's discoveries in the language of calculus, and applied similar methods to heat, light, sound, waves, fluids, electricity, and magnetism. Mathematical physics was born.

The most important message from the *Principia* was that instead of concentrating on how nature behaves, we should seek the deeper laws that govern its behaviour. Knowing the laws, we can deduce the behaviour, gaining power over our environment and reducing uncertainty. Many of those laws take a very specific form: they are differential equations, which express the state of a system at any given moment in terms of the rate at which the state is changing. The equation specifies the laws, or the rules of the game. Its *solution* specifies the behaviour, or how the game plays out, at *all* instants of time: past, present, and future. Armed with Newton's equations, astronomers could predict, with great accuracy, the motion of the Moon and planets, the timing of eclipses, and the orbits of asteroids.

The uncertain and erratic motions of the heavens, guided by the whims of gods, were replaced by a vast cosmic clockwork machine, whose actions were entirely determined by its structure and its mode of operation.

Humanity had learned to predict the unpredictable.

In 1812 Laplace, in his *Essai philosophique sur les probabilités* (Philosophical Essay on Probabilities) asserted that in principle the universe is entirely deterministic. If a sufficiently intelligent being knew the present state of every particle in the universe, it would be able to deduce the entire course of events, both past and future, with exquisite detail. 'For such an intellect,' he wrote, 'nothing would be uncertain, and the future just like the past would be present before its eyes.' This view is parodied in Douglas Adams's *The Hitchhiker's Guide to the Galaxy* as the supercomputer Deep Thought, which ponders the Ultimate Question of Life, the Universe, and Everything, and after seven and a half million years gives the answer: 42.

For the astronomers of his era, Laplace was pretty much right. Deep Thought would have obtained excellent answers, as its real-world counterparts do today. But when astronomers started to ask more difficult questions, it became clear that although Laplace might be right in principle, there was a loophole. Sometimes predicting the future of some system, even just a few days ahead, required impossibly accurate data on the system's state *now*. This effect is called chaos, and it has totally changed our view of the relation between determinism and predictability. We can know the laws of a deterministic system perfectly but still be unable to predict it. Paradoxically, the problem doesn't arise from the future. It's because we can't know the present accurately enough.

SOME DIFFERENTIAL EQUATIONS ARE EASY to solve, with well-behaved answers. These are the linear equations, which roughly means that effects are proportional to causes. Such equations often apply to nature when changes are small, and the early mathematical physicists accepted this restriction in order to make headway. Nonlinear equations are difficult – often impossible to solve before fast computers appeared – but they're usually better models of nature. In the late 19th century, the French mathematician Henri Poincaré

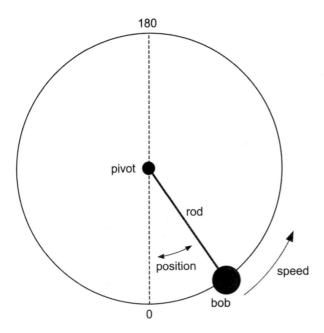

A pendulum, and the two variables specifying its state: *position*, measured as an angle anticlockwise, and *speed*, also measured anticlockwise (angular velocity).

introduced a new way to think about nonlinear differential equations, based on geometry instead of numbers. His idea, the 'qualitative theory of differential equations', triggered a slow revolution in our ability to handle nonlinearity.

To understand what he did, let's look at a simple physical system, the pendulum. The simplest model is a rod with a heavy bob on the end, pivoted at a fixed point and swinging in a vertical plane. The force of gravity pulls the bob downwards, and initially we assume no other forces are acting, not even friction. In a pendulum clock, such as an antique grandfather clock, we all know what happens: the pendulum moves to and fro in a regular manner. (A spring or a weight on a pulley compensates for any energy lost through friction.) Legend has it that Galileo got the idea for a pendulum clock when he was watching a lamp swinging in a church and noticed that the timing was the same whatever angle it swung through. A linear model confirms this, as long as the swings are very small, but a more accurate nonlinear model shows that this isn't true for larger swings.

The traditional way to model the motion is to write down a

differential equation based on Newton's laws. The acceleration of the bob depends on how the force of gravity acts in the direction the bob is moving: tangent to the circle at the position of the bob. The speed at any time can be found from the acceleration, and the corresponding position found from that. The dynamical state of the pendulum depends on those two variables: position and speed. For example, if it starts hanging vertically downwards at zero speed, it just stays there, but if the initial speed isn't zero it starts to swing.

Solving the resulting nonlinear model is hard; so hard that to do it exactly you have to invent new mathematical gadgets called elliptic functions. Poincaré's innovation was to think geometrically. The two variables of position and speed are coordinates on a so-called 'state space', which represents all possible combinations of the two variables – all possible dynamic states. Position is an angle; the usual choice is the angle measured anticlockwise from the bottom. Since 360° is the same angle as 0°, this coordinate wraps round into a circle, as shown in the diagram. Speed is really angular velocity, which can be any real number: positive for anticlockwise motion, negative for clockwise. State space (also called phase space for reasons I can't fathom) is therefore a cylinder, infinitely long with circular cross-section. The position along it represents the speed, the angle round it represents the position.

If we start the pendulum swinging at some initial combination of position and speed, which is some point on the cylinder, those two numbers change as time passes, obeying the differential equation. The point moves across the surface of the cylinder, tracing out a curve (occasionally it stays fixed, tracing a single point). This curve is the trajectory of that initial state, and it tells us how the pendulum moves. Different initial states give different curves. If we plot a representative selection of these, we get an elegant diagram, called the phase portrait. In the picture I've sliced the cylinder vertically at 270° and opened it flat to make the geometry clear.

Most of the trajectories are smooth curves. Any that go off the right-hand edge come back in at the left-hand edge, because they join up on the cylinder, so most of the curves close up into loops. These smooth trajectories are all periodic: the pendulum repeats the same motions over and over again, forever. The trajectories surrounding point A are like the grandfather clock; the pendulum swings to and fro,

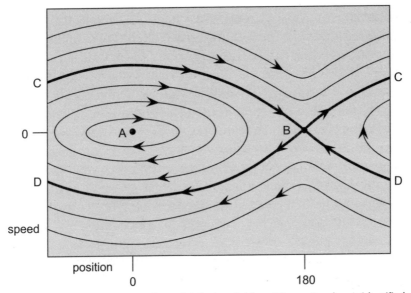

Phase portrait of pendulum. Left- and right-hand sides of the rectangle are identified, because position is an angle. A: centre. B: saddle. C: homoclinic trajectory. D: another homoclinic trajectory.

never passing through the vertical position at 180°. The other trajectories, above and below the thick lines, represent the pendulum swinging round and round like a propeller, either anticlockwise (above the dark line) or clockwise (below).

Point A is the state where the pendulum is stationary, hanging down vertically. Point B is more interesting (and not found in a grandfather clock): the pendulum is stationary, pointing vertically upwards. Theoretically, it can balance there forever, but in practice the upward state is unstable. The slightest disturbance and the pendulum will fall off and head downwards. Point A is stable: slight disturbances just push the pendulum on to a tiny nearby closed curve, so it wobbles a little.

The thick curves C and D are especially interesting. Curve C is what happens if we start the pendulum very close to vertical, and give it a very tiny push so that it rotates anticlockwise. It then swings back up until it's nearly vertical. If we give it just the right push, it rises ever more slowly and tends to the vertical position as time tends to infinity. Run time backwards, and it also approaches the vertical, but from the other side. This trajectory is said to be homoclinic: it limits on the same (*homo*) steady state for both forward and backward infinite time. There's a second homoclinic trajectory D, where the rotation is clockwise.

We've now described all possible trajectories. Two steady states: stable at A, unstable at B. Two kinds of periodic state: grandfather clock and propeller. Two homoclinic trajectories: anticlockwise C and clockwise D. Moreover, the features A, B, C, and D organise all of these together in a consistent package. A lot of information is missing, however; in particular the timing. The diagram doesn't tell us the period of the periodic trajectories, for instance. (The arrows show some timing information: the direction in which the trajectory is traversed as time passes.) However, it takes an infinite amount of time to traverse an entire homoclinic trajectory, because the pendulum moves ever more slowly as it nears the vertical position. So the period of any nearby closed trajectory is very large, and the closer it comes to C or D, the longer the period becomes. This is why Galileo was right for small swings, but not for bigger ones.

A stationary point (or equilibrium) like A is called a centre. One like B is a saddle, and near it the thick curves form a cross shape. Two of them, opposite each other, point towards B; the other two point away from it. These I'll call the in-set and out-set of B. (They're called stable and unstable manifolds in the technical literature, which I think is a bit confusing. The idea is that points on the in-set move towards B, so that's the 'stable' direction. Points on the out-set move away, the 'unstable' direction.)

Point A is surrounded by closed curves. This happens because we ignored friction, so energy is conserved. Each curve corresponds to a particular value of the energy, which is the sum of kinetic energy (related to the speed) and potential energy (arising from gravity and depending on position). If there's a small amount of friction, we have a 'damped' pendulum, and the picture changes. The closed trajectories turn into spirals, and the centre A becomes a sink, meaning all nearby states move towards it. The saddle B remains a saddle, but the out-set C splits into two pieces, both of which spiral towards A. This is a heteroclinic trajectory, connecting B to a different (*hetero*) steady state A. The in-set D also splits, and each half winds round and round the cylinder, never getting near A.

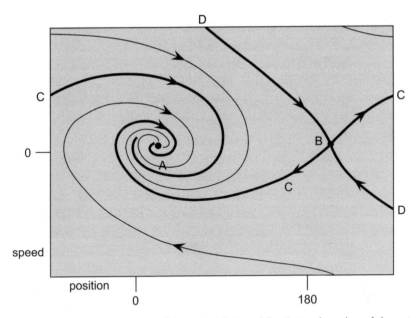

Phase portrait of a damped pendulum. A: sink. B: saddle. C: two branches of the out-set of the saddle, forming heteroclinic connections to the sink. D: two branches of the in-set of the saddle.

These two examples, the frictionless and damped pendulums, illustrate all of the main features of phase portraits when the state space is two-dimensional, that is, the state is determined by two variables. One caveat is that, as well as sinks, there can be sources: stationary points from which trajectories lead outwards. If you reverse all the arrows, A becomes a source. The other is that closed trajectories can still occur when energy isn't conserved, though not in a mechanical model governed by friction. When they do, they're usually isolated – there are no other closed trajectories nearby. Such a closed trajectory occurs, for example, in a standard model of the heartbeat, and it represents a normally beating heart. Any initial point nearby spirals ever closer to the closed trajectory, so the heartbeat is stable.

Poincaré and Ivar Bendixson proved a famous theorem, which basically says that any typical differential equation in two dimensions can have various numbers of sinks, sources, saddles, and closed cycles, which may be separated by homoclinic and heteroclinic trajectories, but nothing else. It's all fairly simple, and we know all the main ingredients. That changes dramatically when there are three or more state variables, as we'll now see.

IN 1961 THE METEOROLOGIST EDWARD LORENZ was working on a simplified model of convection in the atmosphere. He was using a computer to solve his equations numerically, but he had to stop in the middle of a run. So he entered the numbers again by hand to restart the calculation, with some overlap to check all was well. After a time, the results diverged from his previous calculations, and he wondered whether he'd made a mistake when he typed the numbers in again. But when he checked, the numbers were correct. Eventually he discovered that the computer held numbers internally to more digits than it printed out. This tiny difference somehow 'blew up' and affected the digits that it did print out. Lorenz wrote: 'One meteorologist remarked that if the theory were correct, one flap of a seagull's wings could change the course of weather forever.'

The remark was intended as a put-down, but Lorenz was right. The seagull quickly morphed into the more poetic butterfly, and his discovery became known as the 'butterfly effect'. To investigate it, Lorenz applied Poincaré's geometric method. His equations have three variables, so state space is three-dimensional space. The picture shows a typical trajectory, starting from lower right. It rapidly approaches a shape rather like a mask, with the left half pointing out of the page towards us, and the right half pointing away from us. The trajectory spirals around inside one half for a while, then switches to the other half, and keeps doing this. But the timing of the switches is irregular – apparently random – and the trajectory isn't periodic.

If you start somewhere else, you get a different trajectory, but it ends up spiralling around the same mask-like shape. That shape is therefore called an attractor. It looks like two flat surfaces, one for each half, which come together at top centre and merge. However, a basic theorem about differential equations says that trajectories never merge. So the two separate surfaces must lie on top of each other, very close together. However, this implies that the single surface at the bottom actually has two layers. But then the merging surfaces also have two layers, so the single surface at the bottom actually has four layers. And...

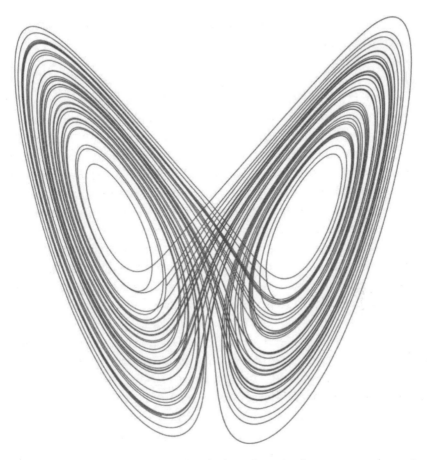

Typical trajectory of the Lorenz equations in three-dimensional space, converging on to a chaotic attractor.

The only way out is that all the apparent surfaces have infinitely many layers, closely sandwiched together in a complicated way. This is one example of a fractal, the name coined by Benoit Mandelbrot for any shape that has detailed structure however much you magnify it.

Lorenz realised that this strange shape explains why his second run on the computer diverged from the first. Imagine two trajectories, starting very close together. They both head off to one half of the attractor, say the left half, and they both spiral round inside it. But as they continue, they start to separate – the spiral paths stay on the attractor, but diverge. When they get near the middle, where the surfaces merge, it's then possible for one to head off to the right half while the other makes a few more spirals round the left half. By the time that one crosses over into the right half, the other one has got so far away that it's moving pretty much independently.

This divergence is what drives the butterfly effect. On this kind of attractor, trajectories that start close together move apart and become essentially independent – even though they're both obeying the same differential equation. This means that you can't predict the future state exactly, because any tiny initial error will grow ever faster until it's comparable in size to the whole attractor. This type of dynamics is called chaos, and it explains why some features of the dynamics look completely random. However, the system is completely deterministic, with no explicit random features in the equations.[36]

Lorenz called this behaviour 'unstable', but we now see it as a new type of stability, associated with an attractor. Informally, an attractor is a region of state space such that any initial condition starting near that region converges towards a trajectory that lies in the region. Unlike the traditional attractors of classical mathematics, points and closed loops, chaotic attractors have more complex topology – they're fractal.

An attractor can be a point or a closed loop, which correspond to a steady state or a periodic one. But in three or more dimensions it can be far more complicated. The attractor itself is a stable object, and the dynamics on it is robust: if the system is subjected to a small disturbance, the trajectory can change dramatically; however, *it still lies on the same attractor.* In fact, almost any trajectory on the attractor explores the whole attractor, in the sense that it eventually gets as near as we wish to any point of the attractor. Over infinite time, nearly all trajectories fill the attractor densely.

Stability of this kind implies that chaotic behaviour is physically feasible, unlike the usual notion of instability, where unstable states are usually not found in reality – for example, a pencil balancing on its point. But in this more general notion of stability, the details are not repeatable, only the overall 'texture'. The technical name for Lorenz's observation reflects this situation: not 'butterfly effect', but 'sensitivity to initial conditions'.

LORENZ'S PAPER BAFFLED MOST METEOROLOGISTS, who were worried that the strange behaviour arose because his model was oversimplified.

It didn't occur to them that if a simple model led to such strange behaviour, a complicated one might be even stranger. Their 'physical intuition' told them that a more realistic model would be better behaved. We'll see in Chapter 11 that they were wrong. Mathematicians didn't notice Lorenz's paper for a long time because they weren't reading meteorology journals. Eventually they did, but only because the American Stephen Smale had been following up an even older hint in the mathematical literature, discovered by Poincaré in 1887–90.

Poincaré had applied his geometric methods to the infamous three-body problem: how does a system of three bodies – such as the Earth, Moon and Sun – behave under Newtonian gravitation? His eventual answer, once he corrected a significant mistake, was that the behaviour can be extraordinarily complex. 'One is struck with the complexity of this figure that I am not even attempting to draw,' he wrote. In the 1960s Smale, the Russian Vladimir Arnold, and colleagues extended Poincaré's approach and developed a systematic and powerful theory of nonlinear dynamical systems, based on topology – a flexible kind of geometry that Poincaré had also pioneered. Topology is about any geometric property that remains unchanged by any continuous deformation, such as when a closed curve forms a knot, or whether some shape falls apart into disconnected pieces. Smale hoped to classify all possible qualitative types of dynamic behaviour. In the end this proved too ambitious, but along the way he discovered chaos in some simple models, and realised that it ought to be very common. Then the mathematicians dug out Lorenz's paper and realised his attractor is another, rather fascinating, example of chaos.

Some of the basic geometry we encountered in the pendulum carries over to more general systems. The state space now has to be multidimensional, with one dimension for each dynamic variable. ('Multidimensional' isn't mysterious; it just means that algebraically you have a long list of variables. But you can apply geometric thinking by analogy with two and three dimensions.) Trajectories are still curves; the phase portrait is a system of curves in a higher-dimensional space. There are steady states, closed trajectories representing periodic states, and homoclinic and heteroclinic connections. There are generalisations of in-sets and out-sets, like those of the saddle point in the pendulum model. The main extra ingredient is that in three or more dimensions, there can be chaotic attractors.

The appearance of fast, powerful computers gave the whole subject a boost, making it much easier to study nonlinear dynamics by approximating the behaviour of the system numerically. This option had always been available in principle, but performing billions or trillions of calculations by hand wasn't a viable option. Now, a machine could do the task, and unlike a human calculator, it didn't make arithmetical mistakes.

Synergy among these three driving forces – topological insight, the needs of applications, and raw computer power – created a revolution in our understanding of nonlinear systems, the natural world, and the world of human concerns. The butterfly effect, in particular, implies that chaotic systems are predictable only for a period of time up to some 'prediction horizon'. After that, the prediction *inevitably* becomes too inaccurate to be useful. For weather, the horizon is a few days. For tides, many months. For the planets of the solar system, tens of millions of years. But if we try to predict where our own planet will be in 200 million years' time, we can be fairly confident that its orbit won't have changed much, but we have no idea whereabouts in that orbit it will be.

However, we can still say a lot about the long-term behaviour in a statistical sense. The mean values of the variables along a trajectory, for example, are the same for all trajectories on the attractor, ignoring rare things such as unstable periodic trajectories, which can coexist inside the attractor. This happens because almost every trajectory explores every region of the attractor, so the mean values depend only on the attractor. The most significant feature of this kind is known as an invariant measure, and we'll need to know about that when discussing the relation between weather and climate in Chapter 11, and speculating on quantum uncertainty in Chapter 16.

We already know what a measure is. It's a generalisation of things like 'area', and it gives a numerical value to suitable subsets of some space, much like a probability distribution does. Here, the relevant space is the attractor. The simplest way to describe the associated measure is to take any dense trajectory on the attractor – one that comes as close as we wish to any point as long as we wait long enough. Given any region of the attractor, we assign a measure to it by following this trajectory for a very long time and counting the proportion of time that it spends inside that region. Let the time become very large, and you've found the measure of that region. Because the trajectory is dense, this in effect defines the probability that a randomly chosen point of the attractor lies in that region.[37]

There are many ways to define a measure on an attractor. The one we want has a special feature: it's dynamically invariant. If we take some region, and let all of its points flow along their trajectories for some specific time, then in effect the entire region flows. Invariance means that as the region flows, its measure stays the same. All of the important statistical features of an attractor can be deduced from the invariant measure. So despite the chaos we can make statistical predictions, giving our best guesses about the future, with estimates of how reliable they are.

DYNAMICAL SYSTEMS, THEIR TOPOLOGICAL FEATURES, AND invariant measures will turn up repeatedly from now on. So we may as well clear up a few more points now, while we're clued in on that topic.

There are two distinct types of differential equation. An ordinary differential equation specifies how a finite number of variables change as time passes. For example, the variables might be the positions of the planets of the solar system. A partial differential equation applies to a quantity that depends on both space and time. It relates rates of change in time to rates of change in space. For example, waves on the ocean have both spatial and temporal structure: they form shapes, and the shapes move. A partial differential equation relates how fast the water moves, at a given location, to how the overall shape is changing. Most of the equations of mathematical physics are partial differential equations.

Today, any system of ordinary differential equations is called a 'dynamical system', and it's convenient to extend that term metaphorically to partial differential equations, which can be viewed as differential equations in infinitely many variables. So, in broad terms, I'll use the term 'dynamical system' for any collection of mathematical rules that determines the future behaviour of some system in terms of its state – the values of its variables – at any instant.

Mathematicians distinguish two basic types of dynamical system:

discrete and continuous. In a discrete system, time ticks in whole numbers, like the second hand of a clock. The rules tell us what the current state will become, one tick into the future. Apply the rule again, and we deduce the state two ticks into the future, and so on. To find out what happens in a million ticks' time, apply the rule a million times. It's obvious that such a system is deterministic: given the initial state, all subsequent states are uniquely determined by the mathematical rule. If the rule is reversible, all past states are also determined.

In a continuous system, time is a continuous variable. The rule becomes a differential equation, specifying how rapidly the variables change at any moment. Subject to technical conditions that are almost always valid, given any initial state, it's possible in principle to deduce the state at any other time, past or future.

THE BUTTERFLY EFFECT IS FAMOUS enough to be lampooned in Terry Pratchett's Discworld books *Interesting Times* and *Feet of Clay* as the quantum weather butterfly. It's less well known that there are many other sources of uncertainty in deterministic dynamics. Suppose there are several attractors. A basic question is: For given initial conditions, which attractor does the system converge towards? The answer depends on the geometry of 'basins of attraction'. The basin of an attractor is the set of initial conditions in state space whose trajectories converge to that attractor. This just restates the question, but the basins divide state space into regions, one for each attractor, and we can work out where these regions are. Often, the basins have simple boundaries, like frontiers between countries on a map. The main uncertainty about the final destination arises only for initial states very near these boundaries. However, the topology of the basins can be much more complicated, creating uncertainty for a wide range of initial conditions.

If state space is a plane, and the shapes of the regions are fairly simple, two of them can share a common boundary curve, but three or more can't. The best they can do is to share a common boundary *point*. But in 1917 Kunizo Yoneyama proved that three sufficiently complicated regions can have a common boundary that doesn't consist

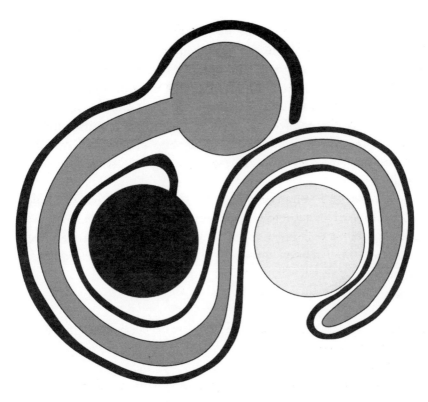

The first few stages in constructing the Lakes of Wada. Each disc puts out ever-finer protuberances that wind between the others. This process continues forever, filling the gaps between the regions.

of isolated points. He credited the idea to his teacher Takeo Wada, and his construction is called the Lakes of Wada.

A dynamical system can have basins of attraction that behave like the Lakes of Wada. An important example arises naturally in numerical analysis, in the Newton–Raphson method. This is a time-honoured numerical scheme for finding the solutions of an algebraic equation by a sequence of successive approximations, making it a discrete dynamical system, in which time moves one tick forward for each iteration. Wada basins also occur in physical systems, such as when light is reflected in four equal spheres that touch each other. The basins correspond to the four openings between spheres, through which the light ray eventually exits.

Riddled basins – like a colander, full of holes – are a more extreme version of Wada basins. Now we know exactly which attractors can occur, but their basins are so intricately intertwined that we have no idea which attractor the system will home in on. In any region of state space, however small, there exist initial points that end up on different attractors. If we knew the initial conditions *exactly*, to infinite precision, we could predict the eventual attractor, but the slightest error makes the final destination unpredictable. The best we can do is to estimate the *probability* of converging to a given attractor.

Riddled basins aren't just mathematical oddities. They occur in many standard and important physical systems. An example is a pendulum driven by a periodically varying force at its pivot, subject to a small amount of friction; the attractors are various periodic states. Judy Kennedy and James Yorke have shown that their basins of attraction are riddled.[38]

11

THE WEATHER FACTORY

A fair day in winter is the mother of a storm.
George Herbert, *Outlandish Proverbs*

FEW THINGS ARE LESS CERTAIN than weather. Yet the underlying physics is very well understood, and we know the equations. So why is weather so unpredictable?

Early pioneers of numerical weather forecasting, hoping to predict the weather by solving the equations, were optimistic. Tides were routinely forecast months ahead. Why not weather? Such hopes were dashed when it became clear that weather is different. Features of the physics make weather unpredictable in the long term, however powerful your computer might be. All computer models are approximate, and unless you're careful, making the equations more realistic can give worse predictions.

Improving the observations might not help much, either. Forecasting is an initial value problem: given the current state of the atmosphere, solve the equations to predict what it will do in the future. But if the dynamics is chaotic, even the smallest error in measuring the current state of the atmosphere blows up exponentially, and the forecast becomes worthless. As Lorenz discovered while modelling a tiny piece of weather, chaos prevents accurate forecasts beyond a specific prediction horizon. For genuine weather this is a few days, even for the most realistic models used by meteorologists.

IN 1922 LEWIS FRY RICHARDSON had a vision of the future, which he made public in *Weather Prediction by Numerical Process*. He had derived a set of mathematical equations for the state of the atmosphere, based on the underlying physical principles, and he proposed using them to make weather forecasts. You just had to input today's data and solve the equations to predict tomorrow's weather. He envisaged what he called a 'weather factory': a huge building full of computers (the word then meant 'people who perform calculations') carrying out some gigantic computation under the guidance of the boss. This august personage would be 'like the conductor of an orchestra in which the instruments are slide rules and calculating machines. But instead of waving a baton he turns a beam of rosy light upon any region that is running ahead of the rest, and a beam of blue light upon those who are behindhand.'

Many versions of Richardson's weather factory exist today, though not in that form: they're weather-forecasting centres armed with electronic supercomputers, not hundreds of people with mechanical calculators. At the time, the best he could do was to grab a calculator and do the sums himself, slowly and laboriously. He tried his hand at numerical weather 'prediction' by forecasting the weather on 20 May 1910, using meteorological observations at 7.00 a.m. to calculate the weather six hours later. His calculations, which took several days, indicated a substantial rise in pressure. Actually, the pressure hardly changed at all.

Pioneering work is always clumsy, and it later turned out that Richardson's strategy was much better than the result suggests. He got the equations and the sums right, but his tactics were faulty, because realistic equations for the atmosphere are numerically unstable. When you convert the equations into digital form, you don't calculate quantities like pressure at every point: only at the corners of a grid. The numerical calculation takes the values at these grid points, and updates them over a very small period of time using an approximation to the true physical rules. The changes in variables like pressure, which determine the weather, are slow and occur on large scales. But the atmosphere can also support sound waves, which are rapid small changes in pressure, and so can the model equations. Sound wave solutions in the computer model can resonate with the grid, and blow up, swamping the actual weather.

Meteorologist Peter Lynch discovered that if modern smoothing methods are used to damp down the sound waves, then Richardson's predictions would have been correct.[39] Sometimes we can improve weather forecasting by making the model equations less realistic.

THE BUTTERFLY EFFECT OCCURS IN mathematical models, but does it happen in the real world? Surely a butterfly can't possibly cause a hurricane. The flap adds a tiny amount of energy to the atmosphere, whereas a hurricane is hugely energetic. Isn't energy conserved? So it is. Mathematically, the flap doesn't create a hurricane out of nothing. Its effect cascades, triggering a rearrangement of weather patterns, small and localised at first, but spreading rapidly, until the entire global weather is markedly different. The energy of the hurricane was there all along, but it was redeployed by the flap. So energy conservation is no obstacle.

Historically, chaos was a surprise because it doesn't show up in equations that are simple enough to solve with a formula. But from Poincaré's geometric viewpoint, chaos is just as reasonable, and just as common, as regular forms of behaviour like steady states and periodic cycles. If some region of state space is stretched locally, but confined to a bounded region, the butterfly effect is inevitable. This can't happen in two dimensions, but it's easy in three or more. Chaotic behaviour may seem exotic, but it's actually quite common in physical systems; in particular, it's why many mixing processes work. But deciding whether it occurs for real weather is trickier. We can't run the weather a second time with the entire planet unchanged except for one butterfly flap, but tests on simpler fluid systems support the view that in principle, real weather is sensitive to initial conditions. Lorenz's critics were wrong; the butterfly effect isn't merely a defect of an oversimplified model.

This discovery has changed the way weather forecasts are computed and presented. The original idea was that the equations are deterministic, so the way to get good long-range forecasts is to improve the accuracy of observations and the numerical methods used to project the current data into the future. Chaos changes all that. Instead, the numerical weather prediction community turned to probabilistic methods, which provide a range of forecasts and an estimate of their accuracy. In practice only the most likely forecast is presented on television or websites, but it's often accompanied by an assessment of how likely it is, such as '25% chance of rain'.

The basic technique here is called ensemble forecasting. 'Ensemble' is a fancy word that physicists use for what mathematicians would call a set. (The term seems to have arisen in thermodynamics.) You make a whole collection of forecasts, not just one. You don't do this the way the 19th-century astronomers did, by making repeated observations of the current state of the atmosphere. Instead, you acquire one set of observational data, and run your ten-day forecast software. Then you make a small random change to the data, and run the software again. Repeat, say 50 times. This gives you 50 samples of what the prediction would have been if you'd based them on the randomly changed observations. In effect, you're exploring the range of predictions that can arise from data close to the numbers observed. You can then count how many of the forecasts predict rain somewhere, and that gives you the probability.

In October 1987 the BBC weather forecaster Michael Fish told viewers that someone had phoned the BBC to warn them that a hurricane was approaching Britain. 'If you're watching,' he said, 'don't worry, because there isn't.'[40] He added that high winds were likely, but the worst of them would be confined to Spain and France. That night, the Great Storm of 1987 hit the south-east of England, with winds gusting to 220 kilometres per hour, and sustained speeds of over 130 kilometres per hour in some regions. Fifteen million trees were blown down, roads were blocked, several hundred thousand people had no electrical power, boats including a Sealink ferry were blown ashore, and a bulk carrier capsized. Insurers paid for £2 billion of damage.

Fish's remarks were based on just one forecast, the right-hand map in the first row of the picture. This was all that was available to him at the time. Later, the European Centre for Medium-Range Weather Forecasts ran a retrospective ensemble forecast using the same data, shown in the rest of the figure. About a quarter of the forecasts in the ensemble show a very deep depression, characteristic of a hurricane.

SIMILAR APPROACHES CAN BE USED for more specific questions. An important application is predicting where a hurricane will go once it has formed. Hurricanes are enormously energetic weather systems, which cause huge damage if they hit land, and their paths are surprisingly erratic. Calculating an ensemble of possible paths can give

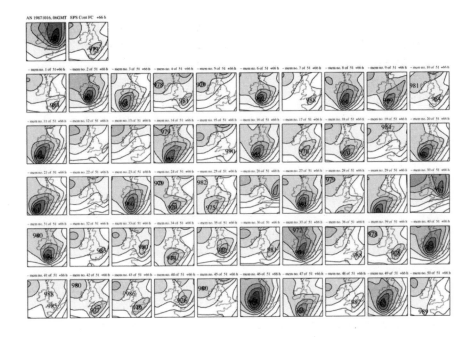

A 66-hour ensemble forecast for 15–16 October 1987. *Top row, second map*: the deterministic forecast. *Top row, first map*: alternative forecast with a deep depression with damaging winds along its southern edge. The remaining 50 panels show other possible outcomes based on small random changes to the initial conditions. On many of them, a deep depression (dark oval) develops.

a ballpark estimate of where and when one is likely to hit, along with statements of the likely size of errors. So cities can plan ahead of time with at least some credible estimate of the danger.

Many detailed mathematical issues are involved in making ensemble forecasts effective. Their accuracy has to be assessed afterwards, to help improve the techniques. The numerical models necessarily approximate the state of a continuous atmosphere by a discrete set of numbers. Various mathematical tricks are used to simplify the computations as far as possible, while retaining the important effects. A big advance was finding tractable ways to include the state of the oceans in the calculations. When several different models are available, a new probabilistic technique opens up: multimodel forecasts. You don't just run many simulations using a single model. You run them using lots of models. This gives some indication not just of their sensitivity to initial conditions, but sensitivity to the assumptions built into the model.

In the early days of numerical weather forecasting, computers didn't exist and hand calculations took days to give a forecast 12 hours ahead. This was useful as proof of concept, and it helped to refine the numerical methods, but it wasn't practical. As computers advanced, meteorologists were able to calculate the weather before it actually happened, but even big official organisations could manage only one forecast per day, using the fastest computers available. Today, there's no difficulty in producing 50 or more forecasts in an hour or two. However, the more data you use, to improve the accuracy of the forecasts, the faster the computer has to be. Multimodel methods stretch their capacity further.

Real weather is subject to other effects too, and sensitivity to initial conditions isn't always the most important cause of unpredictability. In particular, there's not just one butterfly. Small changes to the atmosphere happen all the time, all over the planet. In 1969 Edward Epstein suggested using a statistical model to predict how the mean and variance of atmospheric states change over time. Cecil Leith realised that this approach works only when the probability distributions employed are a good match to the distribution of atmospheric states. You can't just assume a normal distribution, for instance. However, ensemble forecasts based on deterministic but chaotic models soon made these explicitly statistical approaches obsolete.

IT MIGHT BE HOPED THAT, like the small errors in astronomical observations, these innumerable butterfly flaps mostly cancel each other out. But even that, it seems, is too optimistic, a finding that also goes back to Lorenz. The famous butterfly made its entrance in 1972 when he gave a popular lecture with the title: 'Does the flap of a butterfly's wings in Brazil set off a tornado in Texas?' It has long been assumed that this title referred to the 1963 paper, but Tim Palmer, A. Döring, and G. Seregin[41] argue persuasively that he was referring to a different paper of 1969. There he stated that weather systems are unpredictable in a much stronger sense than sensitive dependence on initial conditions.[42]

Lorenz asked how far ahead we can predict a hurricane. The spatial scale of a hurricane is about a thousand kilometres. Within it are medium-scale structures, whose size is more like a hundred kilometres, and within those are cloud systems only a kilometre across. Turbulent vortices in a cloud are a few metres across. Lorenz asked which of these scales is most important for predicting the hurricane. The answer isn't obvious: does the small-scale turbulence 'average out' and have little relevance; or does it gets amplified as time passes and make a big difference?

His answer emphasised three features of this multiscale weather system. First, errors in the large-scale structure double roughly every three days. If this were the only important effect, halving the errors in large-scale observations would extend the range of accurate predictions by three days, opening up the prospect of making good forecasts weeks ahead. However, that's not possible because of the second feature: errors in the fine structure, such as the location of individual clouds, grow much faster, doubling in a hour or so. That's not a problem in its own right, because no one tries to forecast the fine structure, but it becomes a problem thanks to the third feature: errors in the fine structure propagate into the coarser structure. In fact, halving the errors in observations of the fine structure would extend the range of prediction by an hour, not by days. In combination, these three features mean that an accurate forecast two weeks ahead is out of the question.

So far, these statements are about sensitivity to initial conditions. But towards the end Lorenz said something taken directly from the 1969 paper: 'Two states of the system differing initially by a small observational error will evolve into two states differing as greatly as randomly chosen states of the system within a finite time interval, which cannot be lengthened by reducing the amplitude of the initial error.' In other words, there is an absolute limit to the prediction horizon, which can't be extended however accurate your observations are. This is much stronger than the butterfly effect. There, if you could make the observations accurate to enough decimal places, you could make the prediction horizon as long as you please. Lorenz was saying that a week or so is the limit, however accurate your observations are.

Palmer and colleagues go on to show that the situation isn't quite as bad as Lorenz thought. The theoretical issues that create the finite limit don't operate all the time. With ensemble forecasting, it might sometimes be possible to make accurate forecasts two weeks ahead. But doing that would require many more observations, at more closely spaced locations in the atmosphere, and advances in the power and speed of supercomputers.

IT'S POSSIBLE TO MAKE SCIENTIFIC predictions about weather without predicting the weather. A case in point is the common occurrence of blocked weather patterns, in which the atmosphere remains in much the same state for a week or more, before suddenly switching to another long-lived pattern, apparently at random. Examples include the North Atlantic and the Arctic oscillation, in which periods of east–west flow in these regions alternate with periods of north–south flow. Much is known about such states, but transitions between them – which are perhaps their most significant feature – are another matter entirely.

In 1999 Tim Palmer proposed using nonlinear dynamics to improve the prediction of long-term atmospheric behaviour.[43] Daan Crommelin followed up on this idea, providing compelling evidence that blocked atmospheric regimes may be related to the occurrence of a heteroclinic cycle in nonlinear models of large-scale atmospheric dynamics.[44] Recall from Chapter 10 that heteroclinic cycles are sequences of connections between saddle points – equilibrium states that are stable in some directions but unstable in others. Heteroclinic connections create flow patterns that persist for a long time, punctuated by rapid switches from one such pattern to another. This apparently counterintuitive behaviour makes perfect sense in the context of a heteroclinic cycle.

Heteroclinic cycles have an element of unpredictability, but their dynamics is relatively straightforward and much of it is eminently predictable. They're characterised by occasional bursts of activity punctuated by lengthy periods of torpor. The torpid states are predictable: they occur when the system is near an equilibrium. The main element of uncertainty is when the torpor will cease, causing a switch to a new weather pattern.

To detect such cycles in data, Crommelin analyses 'empirical eigenfunctions', or common flow patterns, in the atmosphere. In this technique, the actual flow is approximated by the closest possible combination of a set of independent basic patterns. The complex partial differential equations of meteorology are thereby transformed into a system of ordinary differential equations in a finite number of variables, which represent how strongly the component patterns contribute to the overall flow.

Crommelin tested his theory using data on the northern hemisphere for the period 1948–2000. He found evidence for a common dynamic cycle connecting various regimes of blocked flow in the Atlantic region. The cycle begins with north–south flows over the Pacific and North Atlantic. These merge to form a single east–west Arctic flow. This in turn elongates over Eurasia and the west coast of North America, leading to a predominantly north–south flow. In the second half of the cycle, the flow patterns revert to their original state, but following a different sequence. The complete cycle takes about 20 days. The details suggest that the North Atlantic and Arctic oscillations are related – each may act as a partial trigger for the other.

THE REAL WEATHER FACTORY IS the Sun, which puts heat energy into our atmosphere, oceans, and land. As the Earth spins and the Sun rises and sets, the entire planet goes through a daily cycle of heating and cooling. This drives the weather systems, and leads to many fairly regular large-scale patterns, but the strong nonlinearity of the physical laws makes the detailed effects highly variable. If some influence – a change in the Sun's output, a change in how much heat reflects back into space, a change in how much is retained in the atmosphere – affects the amount of heat energy, the weather patterns will change. If systematic changes go on for too long, the global climate can change.

Scientists have been studying the effects of changes in the planet's heat balance since at least 1824, when Fourier showed that the Earth's atmosphere keeps our planet warm. In 1896 the Swedish scientist Svante Arrhenius analysed the effects on temperature of carbon dioxide (CO_2) in the atmosphere. This is a 'greenhouse gas', helping to trap the Sun's heat. Taking other factors like changes in ice cover (which reflects the Sun's light and heat) into account, Arrhenius calculated that halving the planet's levels of CO_2 could trigger an ice age.

Initially, this theory was mainly of interest to palaeontologists, wondering whether climate change could explain abrupt transitions in the fossil record. But in 1938 the British engineer Guy Callendar collected evidence that both CO_2 and temperature had been rising for the past fifty years, and the question acquired more immediacy. Most scientists either ignored his theory or argued against it, but by the late 1950s several of them were starting to wonder if changes in the concentration of CO_2 were slowly warming the planet. Charles Keeling showed in 1960 that CO_2 levels were definitely rising. A few scientists worried that aerosols might induce cooling and start a new ice age, but papers predicting warming outnumbered those six to one. In 1972 John Sawyer's *Man-made Carbon Dioxide and the 'Greenhouse' Effect* predicted that, by 2000, the expected increase of CO_2 (around 25%) would cause the planet to warm by $0.6°C$. The media continued to emphasise an imminent ice age, but scientists stopped worrying about global cooling, and began to take global warming more seriously.

By 1979 the United States National Research Council was warning that if the rise in CO_2 went unchecked, global temperatures could rise by several degrees. In 1988 the World Meteorological Organisation set up the Intergovernmental Panel on Climate Change to look into the issue, and the world finally began to wake up to impending disaster. Increasingly accurate observations showed that both temperature and CO_2 levels are rising. A 2010 NASA study confirmed Sawyer's forecast. Measurements of the proportions of isotopes (different atomic forms) of carbon in the atmosphere confirm that the main cause of the extra CO_2 is human activity, especially the burning of coal and oil. Global warming became a matter of intense debate. The scientists won the debate years ago, but opponents ('sceptics' or 'denialists' according to taste) continue to challenge the science. Their standard counterarguments sometimes have a naive appeal, but climatologists disproved them all, usually decades ago.

Almost all national governments now accept that global warming is real, that it's dangerous, and that we're the cause. A glaring exception is the current administration in the United States, which seems immune to scientific evidence, and withdrew from the 2015 Paris agreement to limit production of greenhouse gases, for short-term political reasons. After dithering for fifty years, as the delaying tactics of the climate change deniers muddied the waters, the rest of the world

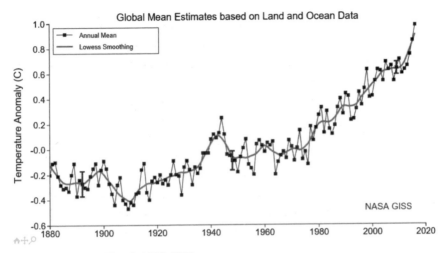

Global Temperature Trends 1880–2020.

is finally taking serious action. Several American states are joining them, despite the noises currently emanating from the White House.

THE 'BEAST FROM THE EAST' was different. A typical British winter storm comes from the west as a region of low air pressure driven by the jet stream, which is a huge vortex of cold air circling the North Pole. Instead, unusually warm conditions in the Arctic in 2018 drove a lot of cold air south, which in turn drove more cold air from Siberia into central Europe and thence to Britain. This also led to storm Emma, and the combination created up to 57 centimetres of snow, temperatures down to −11°C, and killed sixteen people. The unusually cold conditions persisted for over a week, and a lesser event of the same kind occurred a month later.

America, too, has experienced similar events. In 2014 many regions of the USA had a very cold winter; Lake Superior was covered in ice until June, a new record. By July most eastern states other than those bordering the Gulf of Mexico were experiencing weather up to 15°C cooler than normal. At the same time, eastern states had significantly *hotter* weather. The same thing happened again in July 2017, which was the coolest July on record in Indiana and Arkansas, and cooler than usual over most of the eastern USA.

If, as climate scientists confidently assert, human activities are causing the world to get warmer, why do these unprecedentedly cold events keep happening?

The answer is: Because human activities are causing the world to get warmer.

It's not getting warmer by the same amount everywhere. The warming is greatest near the poles: exactly where it can do most damage. Warmer air at the North Pole pushes the jet stream south, and also causes it to weaken, so that it changes position more frequently. In 2014 this effect transported cold air from the poles into the eastern USA. At the same time, the rest of the country received unusually warm air from equatorial regions because the jet stream developed an S-shaped kink. It was one of the ten warmest Julys on record in six western states – Washington, Oregon, Idaho, California, Nevada, and Utah.

Anyone with internet access who genuinely wants to resolve the apparent paradox of unusual cold snaps in a warming world can easily find the explanation, along with the evidence supporting it. You just have to understand the difference between weather and climate.

THE CLIMATE IS ALWAYS CHANGING.

This objection to climate change is an old favourite. US President Donald Trump has repeated it in his tweets about global warming and climate change. Unlike many objections, this one deserves an answer. The answer is: No, the climate is *not* always changing. This may strike you as a silly thing to say, given that some days it's bright and sunny, some days it's raining cats and dogs, and some days everything is buried under a thick layer of snow. *Of course* it's changing all the time! Yes, *it* is. But what's changing is weather, not climate. They're different. Not, perhaps, in everyday language, but in their scientific meaning.

We all understand what weather is. It's what the weather lady tells us about in the television forecast: rain, snow, cloud, wind, sunshine. It refers to what will happen tomorrow, and maybe a few days after. This agrees with the scientific definition of weather: what happens on short timescales – a few hours or a few days. Climate is different. The term is often used more loosely, but what it means is the typical pattern of weather over long timescales – tens of years. The official definition of climate is a 30-year moving average of weather. I'll explain that phrase in a moment, but first, we need to appreciate that averages are subtle.

Suppose that the average temperature over the last 90 days is 16°C. Then there's a heatwave, and it shoots up to 30°C for the next ten days. What happens to the average? It might seem that it ought to get much bigger, but actually it goes up by only 1·4 degrees.[45] Averages don't change much if there are short-term fluctuations. If, on the other hand, the heatwave went on for a total of 90 days, the average temperature over the 180-day period would rise to 23 degrees – halfway between 16 and 30. Long-term fluctuations affect the average more strongly.

A 30-year moving average of temperature is calculated, on any given day, by adding up all the temperatures for every day of the past 30 years, and then dividing by that many days. This number is very stable, and it can change only if temperatures differ from this value over a *very* long period of time – and then only if they tend to move in the same direction (hotter on average, or colder on average). Alternating hot and cold periods to some extent cancel each other out; summer is generally hotter than winter, and the average over the year lies in between. The average over 30 years determines a typical temperature around which everything fluctuates.

This is why climate can't possibly be changing 'all the time'. However dramatically weather is changing today, even if the change is permanent, it will take years for this to have any significant effect on the 30-year average.

Moreover, what we've been considering is just the local climate – in your home town, say. 'Climate change' doesn't refer to your home town. Global climate, the thing that climate scientists tell us is getting warmer, requires the average to be taken not just over a long period of time, but over the entire planet – including the Sahara desert, the Himalayas, the ice-covered polar regions, the Siberian tundra, and the oceans. If it gets colder than usual in Indiana, but warmer than usual in Uzbekistan, the two effects cancel out and the global average stays much the same.

A final remark on terminology. There are some 'natural' (not induced by humans) medium-term effects on climate. The most familiar is El Niño, a warming of the eastern Pacific that occurs naturally every few years. In much of the current literature, 'climate change' is short for 'anthropogenic climate change' – changes in

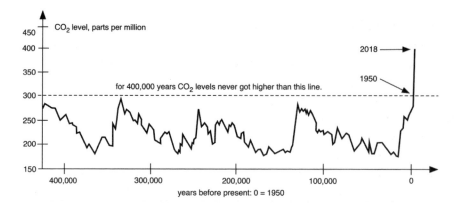

How levels of CO_2 have changed over the past 400,000 years. Until 1950 they never got above 300 parts per million. In 2018 they are 407 parts per million.

climate caused by human activity. Effects like El Niño are accounted for. What's under discussion is changes that aren't, because those happen only if some new factor is affecting the climate.

Many lines of evidence confirm that there is, and it's us. Human activity has pushed the level of CO_2 in the atmosphere above 400 ppm (parts per million). Over most of the previous 800,000 years, it ranged between 170 ppm and 290 ppm (the picture shows the last 400,000 years).[46] Everything above 300 ppm has occurred since the industrial revolution. We know from basic physics that increased CO_2 traps more heat. Global temperatures (inferred from ice cores and ocean sediment by very careful measurements) have risen by nearly 1°C in the last 150 years, which is what physical principles imply the increased CO_2 should create.

IF YOU CAN'T PREDICT THE weather a week ahead, how can you possibly predict the climate in twenty years' time?

This would be a devastating argument if weather and climate were the same thing, but they're not. Some features of a system can be predictable even when others aren't. Remember asteroid Apophis in Chapter 1? The law of gravity tells us that it has a definite chance of hitting the Earth either in 2029 or in 2036, but we don't know if either of those years will actually see a collision. However, we can be absolutely certain that if there is one in either year, it will happen on 13

April. Our ability to predict something depends on what it is, and what we know about it ahead of time. Some features of a system may be predictable with considerable confidence, even though other features are totally unpredictable. Quetelet observed that attributes of individuals are unpredictable, but when averaged over a population, they can often be predicted with considerable accuracy.

For climate, all we have to model are the slow, long-term changes to a 30-year moving average. For climate *change*, we must also model how those long-term relate to changing circumstances – such as the quantity of greenhouse gases added by human activity. This is no easy task, because the climate system is very complex, and its response is subtle. Warmer temperatures may create more clouds, which reflect more incoming heat from the sun. They also melt more ice, and the dark water that replaces it may reflect *less* incoming heat from the Sun. Vegetation grows differently when CO_2 levels are higher, which might remove more CO_2, so the warming might to some extent be self-correcting. (The latest results suggest that initially the rise in CO_2 leads to more vegetation, but unfortunately, this effect disappears after a decade or so.) The models have to do their best to build in any important effects of this kind.

Climate models (of which there are many) work directly with climate data, not with finer-scale weather data. They range from the very simple to the inordinately complex. Complex models are more 'realistic', incorporating more of the known physics of atmospheric flow and other relevant factors. Simple models focus on a limited number of factors; their advantage is that they are easier to understand and to compute. Excessive realism (hence complexity) isn't always better, but it's always computationally more intensive. The aim of a good mathematical model is to retain all important features while leaving out irrelevant complications. As Einstein allegedly said: '[It] should be made as simple as possible, but not simpler.'[47]

To give you a general feel for what's involved, I'll discuss one type of climate model in general terms. Its objective is to understand how the average global temperature changes over time, past and future. In principle this is an accounting process: the extra heat energy over a year is the difference between the energy income – mainly from the Sun, but we might include CO_2 emitted by volcanoes, say – and the energy expenses: heat radiated or reflected away from the planet. If income exceeds expenditure, the planet will get hotter; if the other way round, it will get cooler. We also need to take account of 'savings' – heat stored somewhere, perhaps temporarily. It's a form of expenditure, but one that may come back to haunt us if we realise it as income.

The income is (fairly) straightforward. The vast bulk of it comes from the Sun, we know how much energy the Sun radiates, and we can calculate how much of it hits the Earth. The expenditure is what causes the main headaches. Every hot body radiates heat, and the physical laws for that process are known. This is where greenhouse gases (mainly carbon dioxide and methane, but also nitrous oxide) come in, because they trap heat by the 'greenhouse effect', so that less of it is radiated away. Reflection is trickier because heat is reflected by ice (which is white, so reflects more light and also more heat), clouds, and so on. Clouds are very intricate shapes, hard to model. Moreover, heat can be absorbed; the oceans in particular provide a massive heat sink (savings). It may later re-emerge (taking some of the savings back out of the account).

The mathematical model takes all of these factors into consideration, and writes down equations for how the heat flows between Sun, planet, atmosphere, and oceans. Then we solve them – a job for the computer. All such models are approximations; all make assumptions about which factors are important and which can be ignored; all can be attacked because of these features. Certainly no single model should be taken as gospel. Sceptics make hay with these quibbles, but the central point is simple: *all* of the models predict that human production of excessive quantities of CO_2, created by burning fossil fuels and stored carbon such as forests, have raised and will raise the temperature of the entire planet, compared with what it would have been without our activities. All that extra energy drives weather patterns to new extremes, averaging out as changes in climate.

The precise rate of increase, and how it will persist, isn't certain, but all of the models agree that it will be several degrees Celsius over the next few decades. Observations show that it has already increased by $0.85°C$ since 1880. In fact, it has been increasing ever since the start of the industrial revolution, when manufacturing industries began consuming fossil fuels in large quantities. The rate of increase has been accelerating, with recent years warming twice as fast as earlier ones. It slowed over the past decade, partly because many countries are finally taking steps to cut back on fossil fuel use, but also because of the global recession caused by the banking collapse in 2008. Now it's accelerating again.

Perhaps the biggest threat is rising sea levels, caused by melting ice and the expansion of the water in the ocean as it warms. But there are many others: loss of sea ice, melting of the permafrost releasing methane (an even more potent greenhouse gas), changes to the geographical distribution of disease-bearing insects, and so on. You could write an entire book about the adverse effects, and many people have, so I won't. The effects on the environment can't be predicted decades ahead with perfect accuracy, as sceptics unreasonably demand, but a great variety of adverse effects are already happening. All of the models predict disasters on a vast scale. The only thing open to debate is how vast and how disastrous.

ONE DEGREE IN A CENTURY is *hardly bad news, is it?*

In many parts of the world the locals would be delighted if the climate got one degree warmer. Few of us would be greatly concerned if that's all that happened. But let's put our brains in gear. If it were as simple as that, climate scientists are intelligent enough to have noticed. They wouldn't be worried; they wouldn't even be issuing warnings. So – like most things in life, and everything in nonlinear dynamics – it's not that simple.

One degree doesn't sound like much, but in the past century we humans have managed to warm up *the entire surface of our planet* by that amount. I simplify: the atmosphere, oceans, and land warm by different amounts. Quibbles aside, the amount of *energy* needed to achieve such a widespread change is enormous. And that's the first problem: we're putting vast quantities of extra energy into the nonlinear dynamical system that constitutes the planet's climate.

Don't believe people who tell you it's arrogant to imagine humans could have such a huge effect. One man can burn down a forest – that's a lot of CO_2. We may be tiny, but there's a lot of us, and we have a lot of machinery, much of it emitting CO_2. Measurements show clearly that we *have* had that effect: look at the graph on page 150. We've had similarly extensive effects in other ways. We've belatedly discovered that throwing rubbish into rivers and seas has distributed enormous amounts of plastic waste throughout the oceans. Moreover, that's undeniably our fault. Nothing else makes plastics. Since we can damage the ocean food chain just by dumping our trash, it's hardly arrogant to suggest we may be damaging the environment by churning out enormous quantities of greenhouse gases: around 120 times as much as all the volcanoes on Earth (including those underwater) put together.[48]

One degree may sound harmless, but a vast planet-wide increase in energy doesn't. When you pump extra energy into a nonlinear dynamical system, it doesn't just change its behaviour by the same small amount everywhere. It changes how violently the system fluctuates – how rapidly it changes, how much it changes, and how irregularly it changes. Summer temperatures don't become uniformly one degree warmer. They fluctuate outside their historical range, causing heatwaves and cold snaps. Any individual event of this kind might be comparable to one that occurred decades ago, but you can spot that something has changed when extreme events become considerably more frequent. Global warming has now reached a stage at which heatwaves are happening that have *no* historical counterpart: the USA in 2010 and again in 2011, 2012, and 2013; Southwest Asia in 2011; Australia in 2012–13; Europe in 2015; China and Iran in 2017. In July 2016 Kuwait reached 54°C and Basra in Iraq reached 53.9°C. These are (to date) the highest temperatures ever recorded on Earth except in Death Valley.

In the first half of 2018, many parts of the world experienced severe heatwaves. Britain had a prolonged drought, damaging growing crops. Sweden suffered from wildfires. Algeria recorded Africa's highest ever temperature of 51.3°C. At least thirty people died from heat exposure when the temperature in Japan exceeded 40°C. The Australian state of New South Wales suffered its worst drought on record, causing water shortages, crop failures, and a lack of food for livestock. California was devastated by the Mendocino Complex wildfire, burning more than a quarter of a million acres of land. As each year passes, records are being broken. The same goes for floods, storms, blizzards – choose your extreme event, it's getting more so.

Extra energy in the weather system also changes the way the air flows. What seems to be happening now, for instance, is that the Arctic is warming much more than other parts of the planet. That alters the flow of the polar vortex, cold winds that flow round and round the pole in high latitudes. It weakens the flow, so the cold air wanders around a lot more, and in particular it comes further south. It can also drive the entire stream of circulating cold air further south, which is what happened in early 2018. It may seem paradoxical that global warming can cause a much colder winter than normal, but it's not. It's what happens when you disturb a nonlinear system by giving it more energy. When Europe was being brought to a standstill by temperatures five degrees lower than normal, the Arctic was basking in temperatures twenty degrees higher than normal. Essentially, our excessive production of CO_2 caused the Arctic to export its cold air *to us*.

THE SAME THING IS HAPPENING at the opposite end of the planet, and it could be even worse. Antarctica contains far more ice than the Arctic. It used to be thought that the Antarctic was melting more slowly than the Arctic. It turned out that it's melting faster, but hidden away at the bases of the coastal ice sheets, deep underwater. This is very bad news, because the ice sheets might destabilise, and all that ice will then flow into the oceans. Huge chunks of ice shelves are breaking off already.

Melting polar ice caps are a truly global problem, because the oceans will rise as the extra water runs into them. Current estimates indicate that if all the ice in the Arctic and Antarctic melted, sea levels would rise by 80 metres or more. That won't happen for a long time yet, but a rise of 2 metres is already thought to be unavoidable, whatever we do. Those figures would be lower if the warming were limited to just one degree. But it's not. It wouldn't be even if the *mean* warming were just one degree – which is what's happened since the industrial revolution. Why not? Because the warming is *not* uniform. At the poles, exactly the places we'd like warming to be as small as possible, the temperature rise is much greater than in more temperate latitudes. The Arctic has warmed by about 5°C, on average. It used to be thought that the Antarctic was less of a threat, since it seemed to be warming less, but that was before scientists looked underwater.

Accurate data on ice loss, when combined with data for new ice accumulating on the surface, are vital for understanding the Antarctic contribution to sea-level rise. The ice sheet mass balance inter-comparison exercise (IMBIE), led by Andrew Shepherd and Erik Ivins, is an international team of polar scientists that provides estimates of the rise in sea level caused by melting ice sheets. Its 2018 report,[49] combining results from 24 independent studies, shows that that between 1992 and 2017 the Antarctic ice sheet lost 2720 ± 1390 billion tonnes of ice, increasing the average global sea level by $7 \cdot 6 \pm 3 \cdot 9$ millimetres ($\pm$ errors represent one standard deviation). In west Antarctica the rate of ice loss, mainly from melting ice shelves at the edge of the continent, has tripled in the last 25 years, from 53 ± 29 billion to 159 ± 26 billion tonnes per year.

Stephen Rintoul and colleagues[50] have investigated two possible scenarios for the Antarctica of 2070. If current trends in emissions continue unchecked – as they will unless we take drastic action on a global scale – then relative to a 1900 baseline, mean global land temperatures will be $3 \cdot 5°C$ higher, well above the $1 \cdot 5 - 2°C$ limit adopted in the Paris agreement. Melting Antarctic ice will have caused 27 centimetres of sea-level rise, over and above rises from other causes. The Southern Ocean will be $1 \cdot 9°C$ warmer, and in summer, 43% of Antarctic sea ice will be lost. The number of invasions by 'alien' species will have increased tenfold; the ecosystem will have changed from current species such as penguins and krill to crabs and salps, a form of plankton.

As the ice melts, it causes a vicious circle of positive feedback, making the problem worse. Fresh ice is white, and reflects some of the Sun's heat back into space. As sea ice melts, the white ice becomes dark water, which absorbs more heat and reflects less. So the region warms even faster. On the glaciers of Greenland, melting ice makes the white glaciers dirty, so they melt faster too. In Siberia and northern Canada, where the ground used to be permafrost – frozen throughout the year – the frost is no longer so perma. That is, the permafrost is melting. Trapped inside it are huge amounts of methane, created by rotting vegetation, and – wouldn't you just know it – methane is a greenhouse gas, and a far more potent one than CO_2.

Not to mention methane hydrate, a solid similar to ice, in which methane molecules are trapped inside a crystalline structure of water. There are gigantic deposits of methane hydrate in the shallower regions of the continental shelf, worldwide. They are estimated to contain the equivalent of three trillion tonnes of CO_2, about a hundred years' worth of current human production. If those deposits start to melt...

You think one degree in a century is hardly bad news? Think again.

It's not all doom and gloom if we can get our act together. If emissions can be kept low, the global temperature rise will be 0·9°C, contributing 6 centimetres to sea-level rise. The Southern Ocean will be 0·7°C warmer. In summer, 12% of Antarctic sea ice will be lost. The ecosystem there will remain as it is now. With the ratification of the Paris agreement by 196 countries, and current rapid improvements to the efficiency and cost of renewable energy sources, the lower emissions scenario is entirely feasible. It's unfortunate that the United States has decided to withdraw from the agreement in order to revive its coal-mining industry, a decision that becomes effective in 2020. However, such a revival is unlikely for economic reasons, whatever delusions the current administration entertains. There's all to play for, but the world can't afford to continue silly political delaying tactics for another fifty years.

NONLINEAR DYNAMICS CAN GIVE US a useful perspective on weather, climate, how they're related, and how they change. They're governed by partial differential equations, so the language of dynamic systems is appropriate. I'll use it as a metaphor to sketch one way of thinking about the mathematics.

The state space for the atmosphere at any given location consists of all possible combinations of temperature, pressure, humidity, and so on. Each point in that space represents one possible set of observations of weather. As time passes, the point moves, tracing out a trajectory. We can read off the weather by following the moving point and seeing which regions of state space it passes through. Different short-term sequences of weather data give different trajectories, all lying on the same (chaotic) attractor.

The difference between weather and climate is that weather is a single path through an attractor, whereas climate is the *whole* attractor. In an unchanging climate, there can be many paths through the same attractor, but in the long term they all have similar statistics. The same events happen with the same overall frequency. Weather changes all the time, because the dynamics leads to many different paths on the same attractor. But *climate* shouldn't, unless something way out of the ordinary is happening. Climate change occurs only when the attractor changes. The greater the change to the attractor, the more dramatic the change in climate is likely to be.

A similar image applies to the entire global weather system. The state space becomes an infinite-dimensional function space, because the variables depend on location on the globe, but the same distinction holds: a weather pattern is a trajectory on an attractor, but climate is the whole attractor. This description is metaphorical, because there's no way to observe the attractor as a whole; we'd need trillions of records of global weather patterns. But we can detect something less detailed: the probability that the weather is in a given state. This is related to the invariant measure on the attractor, so if the probability changes, the attractor must have changed. That's why 30-year statistical averages define climate, why we can use them to monitor whether it's changing, and why we can be confident that it is.

Even when the climate has changed, most weather may still look much like the weather before it changed. This is one reason why we may not notice that the climate has changed. We're like the proverbial frog in a saucepan, gradually being brought to the boil but not jumping out because it's getting warmer so slowly that we don't notice. However, the climate scientists noticed sixty-plus years ago, and what they've realised, and documented, and tested, and tried very hard to *disprove*, without success, is that extreme weather events, all over the world, are becoming more common. The plain fact is: *they are*. Just as global warming predicts.

This is the main reason why scientists nowadays talk of climate change instead of global warming. The phrase describes the *effects* more accurately. The cause remains as before: the planet is heating up because humans are producing excess greenhouse gases.

CHANGES IN THE CLIMATE ATTRACTOR are worrying, because even a small change can have large adverse effects. The general point I want to make applies to all extreme weather events, but let's settle on floods as a convenient choice. Engineers designing flood defences have the useful notion of a ten-year flood, or a fifty-year flood, or a hundred-year

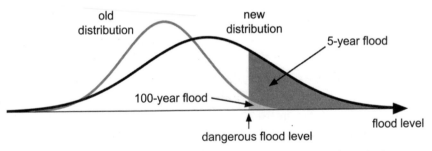

How the 100-year flood can become a 5-year flood when the mean and standard deviation increase.

flood. That's a flood to a level that occurs on average once every ten, fifty, or a hundred years. High levels of flooding are rare, and defending against them is expensive. At some point, the cost outweighs the likely damage. Let's assume that level is the hundred-year flood.

This is all very well as long as the statistics of flood levels remains the same. But what happens if the statistics change? If the mean level of floods increases, higher levels become more likely. The same goes if fluctuations about the mean get bigger: the standard deviation increases. Put the two together, which is what is what all of the extra energy supplied by global warming is likely to create, and they reinforce each other. The picture shows how these effects combine, using a normal distribution for simplicity, but similar reasoning applies to more realistic distributions.

With the old probability distribution, the area under the curve corresponding to dangerous flood levels is the light-grey region, which is small. But when the distribution changes, the area under the curve corresponding to dangerous flood levels includes the dark-grey region as well, which is much larger. Perhaps the new area represents a probability of one such flood every five years. If so, what used to be a hundred-year flood is now a five-year flood. Dangerous floods can happen twenty times as often, and the economic calculations that justified not protecting against the critical level of flooding no longer apply.

In coastal areas, storm surges and rising sea levels add to the dangers posed by heavy and persistent rainfall. Global warming makes all of these causes of flooding much worse. Realistic mathematical models indicate that unless global CO_2 emissions are drastically reduced, then Atlantic City, New Jersey, will soon suffer from chronic flooding.[51] Within thirty years, water levels that currently occur once a century will happen twice a year. The hundred-year flood will become the six-monthly flood, putting homes worth $108 billion at risk. And that's just one coastal city. Currently, 39% of the American population lives in a county adjacent to the coast.

12

REMEDIAL MEASURES

A natural death is where you die by yourself without a doctor's help.
Unknown schoolboy, *Examination script*

IN 1957 A NEW WONDER DRUG became available in Germany. You could get it without a doctor's prescription. Initially it was sold as a cure for anxiety, but later it was recommended to counteract nausea in pregnant women. Its proprietary name was Contergan, but the generic term is thalidomide. After a time, doctors noticed a big increase in the number of babies born with phocomelia – partially formed limbs, leading to death in a proportion of cases – and realised that the drug was responsible. About 10,000 children were affected, with 2000 deaths. Pregnant women were advised not to use the drug, and it was withdrawn in 1959 after it was discovered that prolonged use could cause nerve damage. However, it was later approved for some specific diseases: a form of leprosy and multiple myeloma (cancer of the plasma cells in the blood).

The thalidomide tragedy is a reminder of the uncertain nature of medical treatments. The drug had been extensively tested. The conventional wisdom was that thalidomide would be unable to cross the placental barrier between mother and child, so it could have no effect on the foetus. Nevertheless, the researchers carried out standard tests to detect teratogenic effects – malformations in the foetus – using small laboratory animals. These showed nothing harmful. Later it became clear that humans are unusual in this respect. The medical profession, the pharmaceutical companies that manufacture drugs, the makers of other medical equipment such a replacement hip joints, or just doctors testing different versions of some procedure (such as how best to administer radiotherapy to a cancer patient) have developed methods for testing whether treatments are effective, and reducing risk to patients. As thalidomide shows, these methods aren't foolproof, but they provide a rational way to mitigate the uncertainties. The main tool is statistics, and we can understand several basic statistical concepts and techniques by seeing how they're used in medicine. The methods are constantly being refined as statisticians come up with new ideas.

All such investigations have an ethical dimension, because at some point a new drug, treatment, or treatment protocol has to be tried out on human subjects. In the past, medical experiments were sometimes performed on criminals, unwitting members of the armed forces, the poor and destitute, or slaves, often without their knowledge or consent. Today's ethical standards are more stringent. Unethical experiments still happen, but in most parts of the world they're a rare exception, leading to criminal prosecution if discovered.

The three main types of medical uncertainty concern drugs, equipment, and treatment protocols. All three are developed in the laboratory, and tested before trying them out on humans. The tests sometimes involve animals, opening up new ethical considerations. Animals shouldn't be used unless there's no other way to obtain the required information, and then only under strict safeguards. Some people want animal experiments outlawed altogether.

The later stages of these tests, which are needed to make the drug, device, or treatment regime available for doctors to use on patients, usually involve clinical trials: experiments performed on humans. Government regulators sanction such tests based on an assessment of the risks and potential benefits. Allowing a trial to be carried out need not imply that it's considered safe, so the statistical concept of risk is deeply involved in the entire process.

Different types of trial are used in different circumstances – the whole area is extraordinarily complex – but it's common to start with pilot trials on a small number of people, who may be volunteers or existing patients. Statistically, conclusions based on small samples are less reliable than those involving larger numbers of individuals, but pilot trials give useful information about the risks, leading to improved experimental designs for later trials. For example, if the treatment produces severe side effects, the trial is terminated. If nothing particularly nasty happens, the trial can be extended to larger groups of people, at which stage statistical methods provide a more trustworthy assessment of how effective the treatment concerned is.

If it passes enough tests of this kind, the treatment becomes available for doctors to use, very possibly with restrictions on the type of patient for whom it's considered suitable. Researchers continue to collect data on the outcomes of the treatment, and these data can give increased confidence in its use, or reveal new problems that didn't show up in the original trials.

ASIDE FROM PRACTICAL AND ETHICAL problems, the way a clinical trial is set up centres around two related issues. One is the statistical analysis of the data generated by the trial. The other is the experimental design of the trial: how to structure it so that the data are useful, informative, and as reliable as possible. The techniques chosen for the data analysis affect what data are collected, and how. The experimental design affects the range of data that can be collected, and the reliability of the numbers.

Similar considerations apply to all scientific experiments, so clinicians can borrow techniques from experimental science, and their work also contributes to scientific understanding in general.

Clinical trials have two main objectives: does the treatment work, and is it safe? In practice, neither factor is absolute. Drinking small amounts of water is close to 100% safe (not exactly: you might choke on it), but it won't cure measles. Childhood vaccination against measles is almost 100% effective, but it's not totally safe. Rarely, a child may experience a severe reaction to the vaccine. These are extreme cases, and many treatments are riskier than water and less effective than vaccination. So there may have to be a trade-off. This is where risk enters the picture. The risk associated with an adverse event is the probability it will happen multiplied by the damage it would do.

Even at the design stage, experimenters try to take these factors into account. If there's evidence that people have taken some drug to treat a different condition without suffering severe side effects, the safety issue is to some extent settled. If not, the size of the trial must be kept small, at least until early results come in. One important feature of experimental design is the use of a control group – people who are *not* given the drug or treatment. Comparing the two groups tells us more than just testing the people in the trial on their own. Another important feature is the conditions under which the trial is carried out. Is it structured so that the result is reliable? The thalidomide trial underestimated the potential risk to a foetus; with hindsight, trials on pregnant women should have been given more weight. In practice, the structure of clinical trials evolves as new lessons are learned.

A subtler problem concerns the influence that the experimenter has on the data they collect. Unconscious bias can creep in. Indeed, conscious bias can creep in, as someone sets out to 'prove' some favourite hypothesis and cherry-picks data that fit it. Three important features are common to most clinical trials nowadays. For definiteness, suppose we're testing a new drug. Some subjects will receive the drug; a control group will be given a placebo, a pill that as far as possible appears to be indistinguishable from the drug, but has no significant effect.

The first feature is randomisation. Which patients get the drug, and which get the placebo, should be decided by a random process.

The second is blindness. Subjects should not know whether they got the drug or the placebo. If they knew, they might report their symptoms differently. In a double-blind set-up, the researchers are also unaware of which subjects got the drug or the placebo. This prevents unconscious bias, either in the interpretation of data, its collection, or such things as removing statistical outliers. Going to greater lengths still, a double-dummy design is one that gives each subject both drug and placebo, alternating them.

Thirdly, the use of a placebo as control lets researchers account for the now well known placebo effect, in which patients feel better merely because the doctor has given them a pill. This effect can kick in even when they *know* the pill is a placebo.

The nature of the trial – the disease it aims to cure, or mitigate, and the condition of the subject – may rule out some of these techniques. Giving a placebo to a patient, instead of the drug, can be unethical if it's done without their consent. But with their consent, the trial can't be blind. One way round this, when the trial concerns a new treatment and the aim is to compare it to an existing one that is known to be relatively effective, is to use an 'active control' trial in which some patients get the old treatment and some get the new. You can even tell them what's going on, and get their consent to a random use of the two treatments. The trial can still be blind in such circumstances. Not quite so satisfactory scientifically, perhaps, but ethical considerations override most others.

THE TRADITIONAL STATISTICAL METHODS EMPLOYED in clinical trials were developed in the 1920s at the Rothamsted experimental station, an agricultural research centre. This may seem a far cry from medicine, but similar issues of experimental design and data analysis show up. The most influential figure was Ronald Fisher, who worked at Rothamsted. His *Principles of Experimental Design* laid down many of the central ideas, and included many of the basic statistical tools still in widespread use today. Other pioneers of this period added to the toolkit, among them Karl Pearson and William Gosset (who used the pseudonym 'Student'). They tended to name statistical tests and probability distributions after the symbols representing them, so we now have things like the t-test, chi-squared (χ^2), and the gamma-distribution (Γ).

There are two main ways to analyse statistical data. Parametric statistics models the data using specific families of probability distributions (binomial, normal, and so on) that involve numerical parameters (such as mean and variance). The aim is to find the parameter values for which the model best fits the data, and to estimate the likely range of errors and how significant the fit is. The alternative, non-parametric statistics, avoids explicit models, and relies solely on the data. A histogram, presenting the data without further comment, is a simple example. Parametric methods are better if the fit is very good. Non-parametric ones are more flexible, and don't make assumptions that might be unwarranted. Both types exist in profusion.

Perhaps the most widespread of all these techniques is Fisher's method for testing the significance of data in support (or not) of a scientific hypothesis. This is a parametric method, usually based on the normal distribution. In the 1770s Laplace analysed the sex distribution of nearly half a million births. The data indicated an excess of boys, and he wanted to find out how significant that excess was. He set up a model: equal probabilities of boys and girls, given by a binomial distribution. Then he asked how likely the observed figures are if this model applies. His calculated probability was very small, so he concluded that the observations were highly unlikely to occur if the chances of boys and girls are actually fifty-fifty.

This kind of probability is now known as a p-value, and Fisher formalised the procedure. His method compares two opposite hypotheses. One, the null hypothesis, states that the observations arise by pure chance. The other, the alternative hypothesis, states that they don't, and that's the one we're really interested in. Assuming the null hypothesis, we calculate the probability of obtaining the given data (or data in an appropriate range, since the specific numbers obtained have probability zero). This probability is usually denoted by p, leading to the term p-value.

For example, suppose we count the number of boys and girls in a sample of 1000 births, and we get 526 boys, 474 girls. We want to understand whether the excess of boys is significant. So we formulate the null hypothesis that these figures arise by chance. The alternative hypothesis is that they don't. We're not really interested in the probability of *exactly* these values occurring by chance. We're interested in how extreme the data are: the existence of *more* boys than girls. If the number of boys had been 527, or 528, or any larger number, we would also have had evidence that might indicate an unusual excess. So what matters is the probability of getting 526 or more boys by chance. The appropriate null hypothesis is: this figure *or an even greater excess of boys* arises by chance.

Now we calculate the probability of the null hypothesis happening. At this point it becomes clear that I've missed out a vital ingredient from my statement of the null hypothesis: the theoretical probability distribution that's assumed. Here, it seems reasonable to follow Laplace and choose a binomial distribution with fifty-fifty probabilities of boys and girls, but whichever distribution we choose, it's tacitly built into the null hypothesis. Because we're working with a large number of births, we can approximate Laplace's choice of a binomial distribution by the appropriate normal distribution. The upshot here is that $p = 0.05$, so there's only a 5% probability that such extreme values arise by chance. We therefore, in Fisher's jargon, *reject the null hypothesis* at the 95% level. This means that we're 95% confident that the null hypothesis is wrong, and we accept the alternative hypothesis.

Does that mean we're 95% confident that the observed figures are statistically significant – that they don't arise by chance? No. What it means is hedged about with weasel words: we're 95% confident that the observed figures don't arise by chance, as specified by a fifty-fifty binomial distribution (or the corresponding normal approximation). In other words, we're 95% confident that either the observed figures don't arise by chance, or the assumed distribution is wrong.

One consequence of Fisher's convoluted terminology is that this final phrase can easily be forgotten. If so, the hypothesis we think we're testing isn't quite the same as the alternative hypothesis. The latter comes with extra baggage, the possibility that we chose the wrong statistical model in the first place. In this example, that's not too worrying, because a binomial or normal distribution is very plausible, but there's a tendency to assume a normal distribution by default, even though it's sometimes unsuitable. Students are generally warned about this when they're introduced to the method, but after a while it can fade from view. Even published papers get it wrong.

In recent years, a second problem with p-values has been gaining prominence. This is the difference between statistical significance and clinical significance. For example, a genetic test for the risk of developing cancer might be statistically significant at the 99% level, which sounds pretty good. But in practice it might detect only one extra cancer case in every 100,000 people, while giving a 'false positive' – what appears to be a detection of cancer, but turns out not to be – 1000 times in every 100,000 people. That would render it clinically worthless, despite its high statistical significance.

SOME QUESTIONS ABOUT PROBABILITIES IN medicine can be sorted out using Bayes's theorem. Here's a typical example[52]. A standard method for detecting potential breast cancers in women is to take a mammogram, a low-intensity X-ray image of the breast. The incidence of breast cancer in women aged 40 is about 1%. (Over their entire lifetime it's more like 10%, and rising.) Suppose women of that age are screened using this method. About 80% of women with breast cancer will test positive, and 10% of women without breast cancer will also test positive (a 'false positive'). Suppose that a woman tests positive. What is the probability that she will develop breast cancer?

In 1995 Gerd Gigerenzer and Ulrich Hoffrage discovered that when doctors are asked this question, only 15% of them get the answer right.[53] Most of them go for 70–80%.

We can calculate the probability using Bayes's theorem. Alternatively, we can use the same reasoning as in Chapter 8, as follows. For definiteness, consider a sample of 1000 women in this age group. The size of the sample doesn't matter, because we're looking at proportions. We assume the numbers concerned are exactly as specified by the probabilities – this wouldn't be the case in a real sample, but we're using a hypothetical sample to compute the probabilities, so this assumption is sensible. Of those 1000 women, 10 have cancer, and 8 of those will be picked up by the test. Of the remaining 990 women, 99 will test positive. The total number of positives is therefore 107. Of these, 8 have cancer, a probability of 8/107, which is about 7·5%.

This is about one tenth of what most doctors think when asked to estimate the probability in controlled studies. When dealing with an actual patient, they might take more care than when asked for an estimate off the top of their head. Let's hope so, or better still, equip them with suitable software to save them the trouble. The main error in reasoning is to ignore the false positives, leading to the 80% estimate, or to assume they have a small effect, reducing this figure to 70% or so. That way of thinking fails here because the number of women who don't have cancer is much bigger than the number that do. Even though a false positive is less likely than a genuine positive, the sheer number of women without cancer overwhelms the figures for those who have the condition.

This is yet another instance of fallacious reasoning about conditional probabilities. The doctors are in effect thinking about:

■ The probability that a woman with breast cancer has a positive mammography

when they should be thinking about:

■ The probability that a woman with a positive mammography has breast cancer.

Interestingly, Gigerenzer and Hoffrage showed that doctors estimate this probability more accurately if they're told the figures in a verbal narrative. If 'probability of 1%' is replaced by 'one woman in a hundred' and so on, they mentally visualise something more like the calculation we've just done. Psychological studies show that people are often more able to solve a mathematical or logic question if it's presented as a story, especially in a familiar social setting. Historically, gamblers intuited many basic features of probability long before the mathematicians got to work on it.

IN A MOMENT I'M GOING to take a look at a modern medical trial, which used more sophisticated statistical methods. To prepare for that, I'll start with the methods themselves. Two follow the traditional lines of least squares and Fisher's approach to hypothesis testing, but in less traditional settings. The third is more modern.

Sometimes the only data available are a binary yes/no choice, such as pass/fail in a driving test. You want to find out whether something influences the results, for example whether the number of driving lessons someone takes affects their chances of passing. You can plot the outcome (say 0 for fail, 1 for pass) against the number of hours of lessons. If the outcome were closer to a continuous range, you'd use regression analysis, fit the best straight line, calculate the correlation coefficient, and test how significant it is. With only two data values, however, a straight line model doesn't make a lot of sense.

In 1958 David Cox proposed using logistic regression. A logistic curve is a smooth curve that increases slowly from 0, speeds up, and then slows down again as it approaches 1. The steepness of the rise in the middle, and its location, are two parameters that give a family of such curves. You can think of this curve as a guess about the examiner's *opinion* of the driver on a scale from poor to excellent, or a guess about the actual scores he or she assigned if that's how the test worked. Logistic regression attempts to match this presumed distribution of opinions or scores using only the pass/fail data. It does this by estimating the parameters for which the curve best fits the data, according to whatever definition of 'best fit' we wish. Instead of trying to fit the best straight line, we fit the best logistic curve. The main parameter is usually expressed as the corresponding odds ratio, which gives the relative probabilities of the two possible results.

The second method, Cox regression, was also developed by Cox,

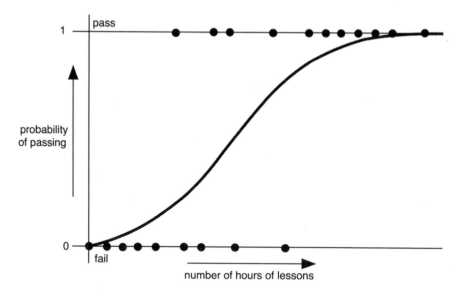

Hypothetical driving-test data (dots) and fitted logistic curve.

and dates from 1972. It's a 'proportional hazards' model, able to deal with events that change over time.[54] For example, does taking some drug make the occurrence of a stroke less likely, and if so, by how much? The hazard rate is a number that tells us how likely a stroke is over a given period of time; doubling the hazard rate halves the average time to a stroke. The underlying statistical model assumes a specific form for the hazard function – how the hazard depends on time. It includes numerical parameters, which model how the hazard function depends on other factors such as medical treatment. The aim is to estimate those parameters and use their values to decide how significantly they affect the likelihood of a stroke, or whatever other outcome is being studied.

The third method is used to estimate the reliability of a statistic calculated from a sample, such as the sample mean. The issue goes back to Laplace, and in astronomy it can be dealt with by measuring the same things many times and applying the central limit theorem. In medical trials and many other areas of science, this may not be possible. In 1979 Bradley Efron suggested a way to proceed without collecting more data, in a paper 'Bootstrap methods: another look at the jackknife'.[55] The first term comes from the saying 'pull yourself up by your bootstraps'; the jackknife was an earlier attempt of a similar kind. Bootstrapping is based on 'resampling' the same data. That is, taking a series of random samples from the existing data, calculating their means (or whatever statistic you're interested in), and finding the resulting distribution of values. If the variance of this resampled distribution is small, the original mean is likely to be close to the true mean of the original population.

For instance, suppose we have data on the heights of a sample of 20 people and want to infer the average height of everyone on the globe. This is a rather small sample so the reliability of the sample mean is questionable. The simplest version of the bootstrap selects 20 of those people at random and calculates the mean for that sample. (The same person can be selected more than once when you resample; statisticians call this 'resampling with replacement'. That way, you don't get the same mean every time.) You resample the data a large number of times, say 10,000. Then you calculate statistics, such as the variance of these 10,000 resampled data points. Or draw a histogram of them. This is easy with computers, but it was impractical before the modern era, so nobody suggested doing it. Strange as it may seem, bootstrapping gives better results than the traditional assumption of a normal distribution or calculating the variance of the original sample.

WE'RE NOW READY TO TAKE a look at a well-designed modern medical trial. I've chosen a research paper from the medical literature, by Alexander Viktorin and coworkers, which appeared in 2018.[56] Their study was about existing drugs, already in widespread use, and they were looking for unintended effects. Specifically, the aim was to examine what happens when a father is using antidepressants at the time a child is conceived. Is there evidence of any harmful effects on the child? They looked at four possibilities: preterm (premature) births, malformations, autism, and intellectual disability.

The study worked with a very large sample of 170,508 children – all children conceived in Sweden between 29 July 2005 and 31 December 2007, as given by the Swedish Medical Birth Register, which covers about 99% of births there. This database includes information that can be used to calculate the date of conception to within a week. Fathers were identified using the Multi-Generation Register provided by Statistics Sweden, which distinguishes between biological and adoptive parents; only biological fathers should be included. If the required data were unavailable, that child was excluded. The regional ethics committee in Stockholm approved the study, and its nature meant that in Swedish law individuals did not have to be asked for consent. As an extra precaution to ensure confidentiality, all data were anonymised: not associated with specific names. The data were collected until 2014, when the child reached the age of 8 or 9.

It turned out that the father had used antidepressants during conception in 3983 cases. A control group of 164,492 children had fathers who had not used them. A third 'negative control group' of 2033 children had fathers who didn't use antidepressants at the time of conception, but did use them later, while the mother was pregnant. (If the drug is harmful, that should show up in the first group, but not in the second. Moreover, we wouldn't *expect* it to show up in the third group either, because the main way a drug or its effects could pass from father to child would be at conception. Testing that expectation is a useful check.)

The study reported that none of the four conditions investigated are caused by the father taking antidepressants during the period around conception. Let's see how the team arrived at these conclusions.

To make the data objective, the investigators used standard clinical classifications to detect and quantify the four adverse conditions. Their statistical analysis used a variety of techniques, each appropriate for the condition and data concerned. For hypothesis testing, the investigators chose the 95% significance level. For two conditions, preterm birth and malformations, the available data were binary: either the child had the condition, or not. An appropriate technique is logistic regression, which provided estimated odds ratios for preterm birth and malformations, quantified using 95% confidence intervals. These define a range of values such that we're 95% confident that the statistic lies inside that range.[57]

The other two conditions, autism spectrum disorder and intellectual disability, are psychiatric disorders. In children, these become more common as the child ages, so the data depend on time. They were corrected for such effects using Cox regression models, providing estimates for hazard ratios. Because data for siblings from the same parents can introduce spurious correlations in the data, the team also used bootstrapping to perform sensitivity analyses to assess the reliability of the statistical results.

Their conclusions supplied statistical evidence to quantify possible associations of antidepressants with the four types of condition. For three conditions, there was no evidence of any association. A second strand was to compare the first group (father used drug during conception) with the third group (father didn't use it during conception, but did during the mother's pregnancy). For the first three conditions again there were no significant differences. For the fourth, intellectual disability, there was a slight difference. If it had indicated a greater risk of intellectual disability for the first group, that might have hinted that the drug was having some effect at conception – the only time when it was likely to affect the eventual foetus. But in fact, the first group had a marginally *lower* risk of intellectual disability than the third.

This is an impressive study. It shows careful design of experiments and correct ethical procedures, and applies a range of statistical techniques that goes well beyond Fisher's style of hypothesis testing. It used traditional ideas such as confidence intervals to indicate the level of confidence in the results, but tailored them to the methods used and the type of data.

13

FINANCIAL FORTUNE-TELLING

Speculators may do no harm as bubbles on a steady stream of enterprise. But the position is serious when enterprise becomes the bubble of a whirlpool of speculation. When the capital development of a country becomes a by-product of the activities of a casino, the job is likely to be ill-done.
John Maynard Keynes, *The General Theory of Employment, Interest, and Money*

ON 15 SEPTEMBER 2008, THE major investment bank Lehman Brothers collapsed. The unthinkable became reality, and the long-term economic boom skidded to an abrupt halt. Simmering anxiety about a specialist area of the American mortgage market boiled over into a full-blown disaster affecting every part of the financial sector. The financial crisis of 2008 threatened to bring down the entire world banking system, a catastrophe averted by governments funnelling huge quantities of taxpayers' money into the very banks that caused it. Its legacy was a global downturn in economic activity of all kinds: The Great Recession. Its malign effects are still widespread a decade later.

I don't want to get tangled up in the causes of the financial crisis, which are complex, varied, and controversial. The general view is that a combination of hubris and greed led to wildly overoptimistic assessments of the value and risk of complicated financial instruments, 'derivatives', which no one really understood. Whatever the cause, the financial crisis offers graphic proof that financial affairs involve deep uncertainty. Before, most of us assumed that the financial world was robust and stable, and that the people in charge of our money were highly trained experts whose extensive experience would engender prudence and a conservative approach to risk. Afterwards, we knew better. There had actually been plenty of previous crises warning that our earlier view was too rosy, but these largely went unnoticed, and if they were noticed, they were dismissed as mistakes that would never be repeated.

There are many kinds of financial institution. There's the everyday kind of bank that you go into to pay in a cheque, or, increasingly, stand outside at a cash machine (ATM), use their banking app, or log in online to transfer money or check that payments have arrived. An investment bank, which lends money for projects, new businesses, and speculative ventures, is very different. The former ought to be risk-free; the latter can't avoid a definite element of risk. In the UK, these two types of bank used to be fenced off from each other. Mortgages used to be provided by Building Societies that were 'mutual', that is, non-profit organisations. Insurance companies stuck to selling insurance, and supermarkets stuck to selling meat and vegetables. Financial deregulation in the 1980s changed all that. Banks piled into mortgage lending, Building Societies abandoned their social role and turned into banks, supermarkets sold insurance. By abolishing allegedly onerous regulations, governments of the time also abolished the firewalls between different types of financial institution. So when a few major banks got into trouble with 'subprime' mortgages,[58] it turned out that everyone else had been making the same mistake, and the crisis spread like wildfire.

That said, financial matters are very hard to predict. The stock market is basically a gambling den – highly organised, useful as a source of finance for business, contributing to the creation of jobs, but ultimately akin to placing a bet on Galloping Girolamo in the 4.30 at Sandown. The currency markets, in which traders exchange dollars for euros, or either for the yen, rouble, or pound, mainly exist in order to make a very small percentage profit on a very big transaction. Just as experienced professional gamblers know the odds, and try to optimise their bets, so professional dealers and traders use their experience to try to keep risk low and profit high. But the stock market is more complex than a horse race, and nowadays traders rely on complicated algorithms – mathematical models implemented on computers. A lot of trading is automated: algorithms make split-second decisions and trade with each other without any human input.

All of these developments have been motivated by the wish to make financial matters more predictable; to reduce uncertainty, and hence reduce risk. The financial crisis came about because too many bankers thought they'd done that. As it turned out, they might as well have been gazing into crystal balls.

IT'S NOT A NEW PROBLEM.

Between 1397 and 1494 the powerful Medici family in Renaissance Italy ran a bank, the biggest and most respectable in the whole of Europe. For a time, it made the Medicis the richest family in Europe. In 1397 Giovanni di Bicci de' Medici split his bank off from his nephew's bank and moved it to Florence. It expanded, with branches in Rome, Venice, and Naples, and then put out tentacles to Geneva, Bruges, London, Pisa, Avignon, Milan, and Lyon. Everything seemed to be going swimmingly under the control of Cosimo de' Medici, until he died in 1464 and was replaced by his son Piero. Behind the scenes, however, the Medicis were spending lavishly: around 17,000 gold florins a year from 1434 to 1471. That's somewhere around 20 to 30 million dollars in today's money.

Hubris begets nemesis, and the inevitable collapse began in the Lyon branch, which had a crooked manager. Then the London branch lent large sums to the current rulers, a risky decision since kings and queens were somewhat ephemeral and notorious for not paying their debts. By 1478 the London branch had failed, with massive losses of 51,533 gold florins. The Bruges branch made the same mistake. According to Niccolò Machiavelli, Piero tried to shore up the finances by calling in debts, which bankrupted several local businesses and annoyed a lot of influential people. Branches kept dying off one by one, and when the Medicis fell from grace in 1494 and lost their political influence, the end was in sight. Even at this stage the Medici bank was the biggest in Europe, but a mob burned the central bank in Florence to the ground and the Lyon branch was subjected to a hostile takeover. The manager there had agreed to too many bad loans, and covered up the disaster by borrowing large amounts from other banks.

It all sounds horribly familiar.

During the dotcom bubble of the 1990s, when investors sold off their holdings in huge profitable industries that actually made goods, and gambled them on what were often half a dozen kids in an attic with a computer and a modem, the Federal Reserve Board chairman Alan Greenspan gave a speech in 1996 castigating the market for its 'irrational exuberance'. No one took any notice, but in 2000 internet stocks plummeted. By 2002 they had lost $5 trillion in market capitalisation.

This, too, had happened before. Many times.

Seventeenth-century Holland was prosperous and confident, making huge profits from trade with the Far East. The tulip, a rare flower from Turkey, became a status symbol and its value soared. 'Tulipomania' exploded, leading to the creation of a specialist tulip exchange. Speculators bought up stock and hid it away to create an artificial scarcity and ramp up prices. A futures market, trading contracts to buy or sell tulip bulbs at a future date, sprang up. By 1623 a sufficiently rare tulip was worth more than an Amsterdam merchant's house. When the bubble burst, it set the Dutch economy back forty years.

In 1711 British entrepreneurs founded the 'Governor and Company of the Merchants of Great Britain, trading to the South Seas and other parts of America, and for the Encouragement of Fishing' – the South Sea Company. The crown granted it a monopoly to trade with South America. Speculators drove the price up tenfold, and people got so carried away that weird spin-off companies were set up. One, famously, had the prospectus: 'For carrying on an undertaking of great advantage, but nobody to know what it is.' Another made square cannonballs. When sanity returned, the market collapsed; ordinary investors lost their life savings, but the big shareholders and directors had jumped ship long before. Eventually Horace Walpole, First Lord of the Treasury, who had sold all his shares at the top of the market, restored order by splitting the debt between the government and the East India Company. The directors were forced to compensate investors, but many of the worst offenders got away with it.

WHEN THE SOUTH SEA BUBBLE burst, Newton, then Master of the Mint and therefore expected to understand high finance, remarked: 'I can calculate the movement of the stars, but not the madness of men.' It took a while before mathematically minded scholars tackled the machinery of the marketplace, and even then they focused on rational decision-making, or at least their best guesses about which behaviours are rational. Economics started to acquire the trappings of a mathematical science in the 19th century. The idea had been brewing for some time, with contributions from people like the German Gottfried Achenwall, often credited with inventing the term 'statistics', and Sir William Petty in England, who wrote about taxation in the mid-1600s. Petty proposed that taxes should be fair, proportionate, regular, and based on accurate statistical data. By 1826 Johann von Thünen was constructing mathematical models of economic systems, such as the use of farmland for agriculture, and developing techniques to analyse them.

Initially the methods were based on algebra and arithmetic, but a new generation of scholars trained in mathematical physics moved in. William Jevons, in *The Principles of Political Economy*, argued that economics 'must be mathematical simply because it deals with quantities'. Collect enough data on how many goods are sold and at what prices, and the mathematical laws underpinning economic transactions would surely be revealed. He pioneered the use of marginal utility: 'As the quantity of any commodity, for instance plain food, which a man has to consume, increases, so the utility or benefit derived from the last portion used decreases in degree.' That is, once you have enough, anything more becomes less useful than it would otherwise have been.

Mathematical economics, in its 'classical' form, can be clearly discerned in the work of Léon Walras and Augustin Cournot, who emphasised the concept of utility: how much a given good is worth to the person buying it. If you're buying a cow, you balance out the cost, including feeding it, against the income from milk and meat. The theory was that the purchaser decides between a variety of possible choices by selecting the one that maximises utility. If you write down plausible formulas for the utility function, which encodes how utility depends on the choice, you can use calculus to find its maximum. Cournot was a mathematician, and in 1838 he developed a model in which two companies compete for the same market, a set-up called a duopoly. Each adjusts its prices according to how much the other is producing, and together they settle into an equilibrium (or steady state) in which they both do as well as they can. The word *equilibrium* comes from the Latin for 'equal balance', and indicates that once such a state is reached, it doesn't change. The reason in this context is that any change will disadvantage one or other company.

Equilibrium dynamics and utility came to dominate mathematical economic thinking, and a major influence in that direction was Walras's attempt to extend such models to the entire economy of a nation, or perhaps even the whole world. This was his theory of general competitive equilibrium. Write down equations describing the choices of the buyer and seller in any transaction, put them all together for *every* transaction on the planet, solve for an equilibrium state, and you've found the best possible choices for everyone. These general equations were too complex to be solved by the methods then available, but they led to two basic principles. Walras's law says that if all markets but one are in equilibrium, so is the final one; the reason is that if that one can vary, it will cause the others to vary as well. The other was *tâtonnement*, a French word ('groping toward') that embodied his view of how real markets attained equilibrium. The market is viewed as an auction, where the auctioneer proposes prices and buyers bid for their desired basket of goods when the price suits their preferences. They are assumed to have such preferences (reservation prices) for every good. One deficiency of this theory is that nobody actually buys anything until all the goods have been auctioned, and no one revises their reservation prices as the auction proceeds. Real markets don't do this; in fact, it's not at all clear that 'equilibrium' is a state that applies meaningfully to real markets. Walras was building a simple deterministic model of a system riddled with uncertainty. His approach persisted because no one could come up with anything better.

Edgeworth, busily tidying up the mathematical formalism of statistics, applied a similar approach to economics in 1881 in *Mathematical Psychics: An Essay on the Application of Mathematics to the Moral Sciences*. New mathematical methods started to appear in the early 20th century. Vilfredo Pareto developed models in which economic agents trade goods with the aim of improving their selection. The system is in equilibrium if it reaches a state where no agent can improve their own selection without making another agent worse off. Such a state is now called a Pareto equilibrium. In 1937 John von Neumann proved that equilibrium states always exist in a suitable class of mathematical models by invoking a powerful theorem from topology, the Brouwer fixed-point theorem. In his set-up, economies can grow in value, and he proved that in equilibrium the growth rate should equal the interest rate. He also developed game theory, a simplified mathematical model of competing agents choosing from a finite range of strategies to maximise their payoff. Later John Nash won the Nobel Prize in Economics[59] for work on equilibria in game theory, closely related to Pareto equilibria.

By the mid-20th century, most of the key features of classical mathematical economics, still widely taught in universities and for many decades the only mathematical approach to economics that was taught anywhere, were firmly in place. Much of the archaic terminology still in use today (market, basket of goods) and the emphasis on growth as a measure of the health of an economy stem from this era. The theory provides a systematic tool for making decisions in an uncertain economic environment, and it works well enough, often enough, to be useful. However, it's becoming increasingly obvious that this type of mathematical model has serious limitations. In particular, the idea that economic agents are perfectly rational beings who know exactly what their utility curve is, and seek to maximise it, fails to match reality. A remarkable feature of the widespread acceptance of classical mathematical economics is that very little of it was ever tested against real data. It was a 'science' with no experimental basis. The great economist John Maynard Keynes wrote: 'Too large a proportion of recent "mathematical" economics are merely concoctions, as imprecise as the initial assumptions they rest on, which allow the author to lose sight of the complexities and interdependencies of the real world in a maze of pretentious and unhelpful symbols.' Towards the end of this chapter we'll take a quick look at some modern proposals for something better.

A VERY DIFFERENT APPROACH TO ONE branch of financial mathematics can be seen in the PhD thesis of Louis Bachelier, which he defended in Paris in 1900. Bachelier was a student of Henri Poincaré, who was probably the leading French mathematician of that period and among the best in the world. The title of the thesis was *Théorie de la spéculation* (Theory of Speculation). This might have been the name of some technical area of mathematics, but Bachelier was referring to speculation in stocks and shares. It wasn't an area to which

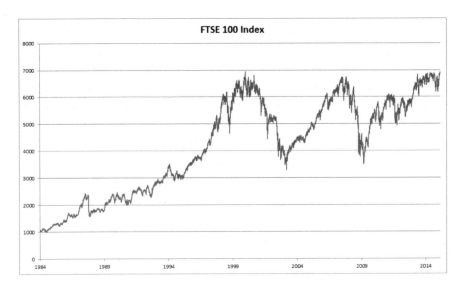

The FTSE 100 index from 1984 to 2014.

mathematics had conventionally been applied, and Bachelier suffered as a result. His mathematics was spectacular in its own right and a major contribution to mathematical physics, where the same ideas apply in a different realisation, but it disappeared without trace until rediscovered decades later. Bachelier pioneered a 'stochastic' approach to financial uncertainty – a technical term for models that have a built-in random element.

Anyone who reads the financial pages of a newspaper, or keeps an eye on the stock market on the web, rapidly discovers that the values of stocks and shares change in an irregular, unpredictable way. The figure shows how the FTSE 100 index (a composite of the values of the top 100 companies on the UK stock market) varied between 1984 and 2014. It looks more like a random walk than a smooth curve. Bachelier took this similarity to heart, and modelled changes in stock prices in terms of a physical process called Brownian motion. In 1827 the Scottish botanist Robert Brown was using a microscope to look at tiny particles trapped in cavities inside pollen grains suspended in water. He noticed that the particles jiggled around randomly, but couldn't explain why. In 1905 Einstein suggested that the particles were colliding with water molecules. He analysed the physics mathematically, and his results convinced many scientists that matter is made of atoms. (It's amazing that this idea was highly controversial in 1900.) Jean Perrin confirmed Einstein's explanation in 1908.

Bachelier used the model of Brownian motion to answer a statistical question about the stock market: How does the expected price – the statistical average – change over time? In more detail, what does the probability density of the price look like, and how does it evolve? The answers provide an estimate of the most likely future price and tells us how much it might fluctuate relative to that price. Bachelier wrote down an equation for the probability density, now called the Chapman–Kolmogorov equation, and solved it, obtaining a normal distribution whose variance (spread) increases linearly as time passes. We now recognise this as the probability density for the diffusion equation, also called the heat equation because that's where it first made its appearance. If you heat a metal saucepan on a cooker, the handle gets hot, even though it's not in direct contact with the heating element. This happens because heat diffuses through the metal. In 1807 Fourier had written down a 'heat equation' governing this process. The same equation applies to other kinds of diffusion, such as an ink drop spreading through a glass of water. Bachelier proved that in a Brownian motion model, the price of an option spreads like heat.

He also developed a second approach, using random walks. If a random walk takes finer and finer steps, ever more rapidly, it approaches Brownian motion. He showed that this way of thinking gave the same result. He then calculated how the price of a 'stock option' should change over time. (A stock option is a contract to buy or sell some commodity at some future date, for a fixed price. These contracts can be bought or sold; whether that's a good idea depends on how the actual price of the commodity moves.) We get the best estimate of the actual future price by understanding how the current price diffuses.

The thesis had a lukewarm reception, probably because of its unusual area of application, but it passed, and was published in a top-quality scientific journal. Bachelier's career was subsequently blighted by a tragic misunderstanding. He continued research on diffusion and related probabilistic topics, becoming a professor at the Sorbonne, but when World War I was declared he joined the army. After the war, and a few temporary academic jobs, he applied for a permanent post at

Dijon. Maurice Gevrey, assessing the candidates, believed he had found a major error in one of Bachelier's papers, and an expert, Paul Lévy, concurred. Bachelier's career lay in ruins. But they had both misunderstood his notation, and there had been no mistake. Bachelier wrote an angry letter pointing this out, to no avail. Eventually Lévy realised Bachelier had been right all along, apologised, and they buried the hatchet. Even so, Lévy was never greatly enamoured of the application to the stock market. His notebook contains a comment on the thesis: 'Too much on finance!'

BACHELIER'S ANALYSIS OF HOW THE value of a stock option changes over time, through random fluctuations, was eventually taken up by mathematical economists and market researchers. The aim was to understand the behaviour of a market in which options, rather than the underlying commodities, are being traded. A fundamental problem is to find a rational way to place a value on an option, that is, a value that everyone concerned could calculate independently using the same rules. This would make it possible to assess the risk involved in any particular transaction, encouraging market activity.

In 1973 Fischer Black and Myron Scholes published 'The pricing of options and corporate liabilities' in the *Journal of Political Economy.* Over the preceding decade they had developed a mathematical formula to determine a rational price for an option. Their experiments in trading with the formula hadn't been terribly successful, and they decided to make the reasoning public. Robert Merton gave a mathematical explanation of their formula, which became known as the Black–Scholes options pricing model. It distinguishes fluctuations in the value of the option from the risk of the underlying commodity, leading to a trading strategy known as delta-hedging: repeatedly buy and sell the underlying commodity in a manner that eliminates the risk associated with the option.

The model is a partial differential equation, closely related to the diffusion equation that Bachelier extracted from Brownian motion. This is the Black–Scholes equation. Its solution in any given circumstance, obtained by numerical methods, gives the optimal price of the option. The existence of a unique 'sensible' price (even though it was based on a particular model, which might not apply in

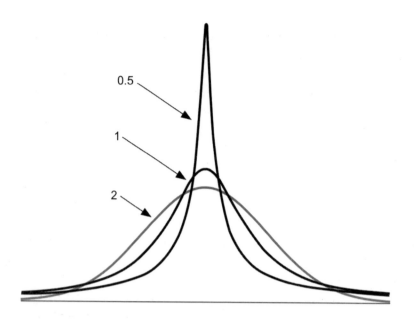

Two fat-tailed distributions (black) and a normal distribution (grey) for comparison. All three are 'stable distributions' involving a parameter whose values are shown (0·5, 1, 2).

reality) was enough to persuade financial institutions to use the equation, and a huge market in options arose.

The mathematical assumptions built into the Black–Scholes equation aren't entirely realistic. An important one is that the probability distribution for the underlying diffusion process is normal, so extreme events are very unlikely. In practice, they're far more common, a phenomenon known as fat tails.[60] The picture shows three members of a four-parameter family of probability distributions called stable distributions, for specific values of a key parameter. When this parameter is 2, we get the normal distribution (grey curve), which doesn't have a fat tail. The other two distributions (black curves) have fat tails: the black curves lie above the grey one near the edges of the picture.

Using a normal distribution to model financial data that actually have fat tails considerably underestimates the risk of extreme events. These events are rare compared with the norm, with or without fat tails, but fat tails make them common enough to be a serious issue. And, of course, it's the extreme events that can lose you large amounts of money. Unexpected shocks, such as sudden political upheavals or the collapse of a major corporation, can make extreme events even more likely than fat-tailed distributions indicate. The dotcom bubble and the 2008 financial crisis both involved this kind of unexpected risk.

Despite these qualms, the Black–Scholes equation was widely used for pragmatic reasons: it was easy to do the sums, and it gave a good approximation to the state of the real market most of the time. The billionaire investor Warren Buffett sounded a cautionary note: 'The Black–Scholes formula has approached the status of holy writ in finance... If the formula is applied to extended time periods, however, it can produce absurd results. In fairness, Black and Scholes almost certainly understood this point well. But their devoted followers may be ignoring whatever caveats the two men attached when they first unveiled the formula.'[61]

More sophisticated and more realistic models were constructed, and developed for more complex financial instruments – 'derivatives'. One cause of the 2008 crisis was a failure to recognise the true risks for some of the most popular derivatives, such as credit default swaps and collateralised debt obligations. Investments that the models claimed to be risk-free weren't.

IT'S NOW BECOMING OBVIOUS THAT traditional mathematical economics, and financial models based on traditional statistical assumptions, are no longer fit for purpose. Less obvious is what we should do about it. I'll take a quick look at two different approaches: a 'bottom-up' analysis which models the actions of individual dealers and traders, and a 'top-down' approach to the overall state of the market, and how to control it so that crashes don't happen. These are a tiny sample from a vast literature.

In the 1980s mathematicians and scientists got very interested in 'complex systems', in which large numbers of individual entities interact through relatively simple rules to produce unexpected 'emergent' behaviour on the level of the overall system. The brain, with its 10 billion neurons, is a real-world example. Each neuron is (fairly) simple, and so are the signals that pass between them, but put enough together in the right way and you get a Beethoven, an Austen, or an Einstein. Model a crowded football stadium as a system of 100,000 individuals, each with its own intentions and capabilities, getting in each other's way or queuing quietly at a ticket-booth, and you can get very realistic predictions of how the crowd flows. For example, dense crowds moving in opposite directions along a corridor can 'interdigitate', forming long parallel lines that alternate their direction of flow. Traditional top-down models of the crowd as a fluid can't reproduce this behaviour.

The state of the stock market has a similar structure: large numbers of traders competing with each other to make a profit. Economists such as W. Brian Arthur began to investigate complexity models of economic and financial systems. One consequence was a style of modelling now called agent-based computational economics (or ACE). The overall structure is quite general. Set up a model in which many agents interact, set up plausible rules for how they do that, run the lot on a computer, and find out what happens. The classical economic assumption of perfect rationality – everyone tries to optimise their own utility – can be replaced by agents with 'bounded rationality', adapting to the state of the market. They do what seems sensible to them, at any given time, based on their own *limited* information about what the market is doing and their guesses about where it's heading. They're not mountaineers climbing towards a distant peak along a trail that they and everyone else can see; they're groping around on the slopes in the mist, heading in a generally upward direction, not even sure if there's a mountain there, but concerned that they might fall off a cliff if they're not careful.

In the mid-1990s, Blake LeBaron examined ACE models of the stock market. Instead of everything settling down to an equilibrium, as classical economics assumes, prices fluctuate as agents observe what's happening and change strategy accordingly. Just like real markets. Some of the models reproduce not just this qualitative behaviour, but the overall statistics of market fluctuations. In the late 1990s, New York's NASDAQ stock exchange was changing from listing prices as fractions (like $23\frac{3}{4}$) to decimals (like 23·7 or perhaps even 23·75). This would allow more accurate pricing, but because prices could move by smaller amounts, it might also affect the strategies the traders used. The stock exchange hired complexity scientists at BiosGroup to develop an ACE model, tuned to produce the correct statistics. It showed that if the price movements were allowed to be *too* small, traders could behave in a way that gave them quick profits while reducing the efficiency of the market. This wasn't a great idea, and NASDAQ took the message on board.

In contrast to this bottom-up philosophy, the Bank of England's Andrew Haldane and ecologist Robert May teamed up in 2011 to suggest that banking can learn lessons from ecology.[62] They observed that an emphasis on the supposed risks (or lack of them) associated with complicated derivatives ignored the collective effect these instruments might have on the overall stability of the entire banking system. Analogously, elephants might be thriving, but if there are too many of them they tear down so many trees that other species suffer. Economists had already shown that the massive growth of hedge funds – economic elephants – can destabilise markets.[63] Haldane and May illustrated their proposal using models deliberately chosen for their simplicity, adapting methods used by ecologists to study interacting species and the stability of ecosystems. One such model is food webs: networks representing which species predate on which. The nodes of the network are individual species; the links between species represent which one feeds on the other, and how. To apply similar ideas to the banking system, each major bank is represented as a node, and what flows between them is money, not food. The analogy isn't bad. The Bank of England and the Federal Reserve Bank of New York developed this structure to explore the consequences of a single bank failure for the banking system as a whole.

Some of the key mathematics of such networks can be captured using a 'mean field approximation', in which every bank is assumed to behave like the overall average. (In Quetelet's terms, assume every bank is the average bank. This isn't so unreasonable because all the big ones copy each other.) Haldane and May examined how the system's behaviour relates to two main parameters: the bank's net worth, and the proportion of assets it holds in interbank loans. The latter involve risk since the loan may not be repaid. If one bank fails, this happens, and the effects propagate through the network.

This model predicts that a bank is most fragile when it's highly active in both retail banking (high street) and investment banking (casino). Since the financial crisis, many governments have belatedly required major banks to split these two activities, so that the failure of the casino doesn't contaminate the high street. The model also includes another way for shocks to propagate through the banking

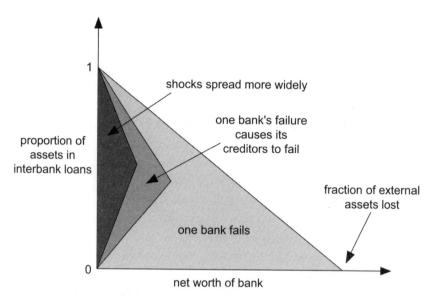

How a bank failure caused by a loss of external assets can spread contagion to creditors or the entire system. Regions show what happens for the corresponding combination of net worth and the proportion of assets held in interbank loans.

system, which was very apparent during the 2008 crisis: banks go into their shells and stop lending to each other. In the jargon, 'funding liquidity shocks' occur. Prasanna Gai and Sujit Kapadia[64] have shown that this behaviour can spread rapidly from bank to bank in a domino effect, and it tends to stay around for a long time unless some central policy gets loans between banks moving again.

Simple top-down models of this kind are useful to inform policy-makers. For example, banks can be required to increase their capital and liquid assets. Traditionally, this form of regulation was seen as a way to stop individual banks taking on too much risk. The ecological model shows that it has a far more important function: preventing a single bank failure cascading through the entire system. Another implication is the need for 'firebreaks' that isolate some parts of the system from others. (These are precisely what were demolished by politically motivated deregulation in the 1980s.) The overall message is that financial regulators should be more like ecologists: concerned with the health of the entire ecosystem, not just individual species.

14

OUR BAYESIAN BRAIN

I used to be indecisive,
But now I'm not so sure.
Message on a T-shirt

IN CHAPTER 2 I ASKED why human beings so easily accept sweeping claims when there's no serious evidence to support them, and why they so readily embrace irrational beliefs even when there's clear evidence against them. Naturally, each of us has his or her own view on which beliefs are, or are not, irrational, but we can all ask those questions about everybody else.

Part of the answer may lie in how our brains evolved, over millions of years, to make rapid decisions about uncertain but life-threatening possibilities. These evolutionary explanations are guesses – it's difficult to see how they could be tested, since brains don't fossilise and there's no way to find out for certain what went on in our ancestors' minds – but they seem plausible. We can be more certain about how modern human brains work, because it's possible to perform experiments that relate brain structure to brain function, and both to genetics.

It would be foolish to underestimate the difficulties of understanding a brain – even that of a fruit fly, let alone a highly complex human being. The fruit fly *Drosophila melanogaster* is a staple of genetic research; its brain contains about 135,000 neurons, linked together by synapses which transmit electrical signals between them. Scientists are currently investigating the structure of this network, known as the fruit fly connectome. At the moment, only two regions out of the 76 main subdivisions of the brain have been mapped. So right now we don't even know the structure of the fruit fly's connectome, let alone how it works. Mathematicians know that even a network of eight or ten neurons can do very puzzling things, because the simplest realistic models of such networks are nonlinear dynamical systems. Networks have special features that aren't typical of general dynamical systems; this may be why nature uses them so much.

The human brain contains about 100 billion neurons, and more than a hundred trillion synapses. Other brain cells may also be involved in its workings, notably glial cells, which occur in roughly the same numbers as neurons, but whose functions remain mysterious.[65] Research is also under way to map the human connectome, not because that will let us simulate a brain, but because it will provide a reliable database for all future brain research.

If mathematicians can't understand a ten-neuron 'brain', what hope is there of understanding a 100-billion-neuron one? Like weather versus climate, it all depends on what questions you ask. Some ten-neuron networks can be understood in considerable detail. Some parts of the brain can be understood even if the whole thing remains bafflingly complex. Some of the general principles upon which the brain is organised can be teased out. In any case, this kind of 'bottom-up' approach, listing the components and how they're linked, and then working our way up towards a description of what the entire system does, isn't the only way to proceed. 'Top-down' analysis, based on the large-scale features of the brain and its behaviour, is the most obvious alternative. In practice, we can mix both together in quite complicated ways. In fact, our understanding of our own brains is growing rapidly, thanks to technological advances that reveal how a network of neurons is connected and what it's doing, and to new mathematical ideas about how such networks behave.

MANY ASPECTS OF BRAIN FUNCTION can be viewed as forms of decision-making. When we look at the external world, our visual system has to figure out which objects it's seeing, guess how they will behave, assess their potential for threat or reward, and make us act in accordance with those assessments. Psychologists, behavioural scientists, and workers in artificial intelligence have come to the conclusion that in some vital respects, the brain seems to function as a Bayesian decision machine. It embodies beliefs about the world, temporarily or permanently wired into its structure, which lead it to make decisions that closely resemble what would emerge from a Bayesian probability model. (Earlier, I said that our intuition for probability is generally rather poor. That's not in conflict with what I've just said, because the inner workings of these probability models aren't consciously accessible.)

The Bayesian view of the brain explains many other features of human attitudes to uncertainty. In particular, it helps to explain why superstitions took root so readily. The main interpretation of Bayesian statistics is that probabilities are *degrees of belief*. When we assess a probability as fifty-fifty, we're effectively saying we're as willing to believe it as we are to disbelieve it. So our brains have evolved to embody beliefs about the world, and these are temporarily or permanently wired into its structure.

It's not just human brains that work this way. Our brain structure goes back into the distant past, to mammalian and even reptilian evolutionary ancestors. Those brains, too, embodied 'beliefs'. Not the kind of beliefs we now articulate verbally, such as 'breaking a mirror brings seven years' bad luck'. Most of our own brain-beliefs aren't like that either. I mean beliefs such as 'if I flick my tongue out this way then I'm more likely to catch a fly', encoded in the wiring of the region of the brain that activates the muscles involved. Human speech added an extra layer to beliefs, making it possible to express them, and more importantly, to pass them on to others.

To set the scene for a simple but informative model, imagine a region of the brain containing a number of neurons. They can be linked together by synapses, which have a 'connection strength'. Some send weak signals, some send strong signals. Some don't exist at all, so they send no signals. The stronger the signal, the greater is the response of the neuron receiving it. We can even put numbers to the strength, which is useful when specifying a mathematical model: in appropriate units, maybe a weak connection has strength 0.2, a strong one 3.5, and a non-existent one 0.

A neuron responds to an incoming signal by producing a rapid change in its electrical state: it 'fires'. This creates a pulse of electricity that can be transmitted to other neurons. Which ones depends on the connections in the network. Incoming signals cause a neuron to fire when they push its state above some threshold value. Moreover, there are two distinct types of signal: excitatory ones, which tend to make the neuron fire, and inhibitory ones, which tend to stop it firing. It's as though the neuron adds up the strengths of the incoming signals, counting excitatory ones as positive and inhibitory ones as negative, and fires only if the total is big enough.

In new-born infants, many neurons are randomly connected, but as time passes some synapses change their strengths. Some may be removed altogether, and new ones can grow. Donald Hebb discovered a form of 'learning' in neural networks, now called Hebbian learning. 'Nerve cells that fire together wire together'. That is, if two neurons fire in approximate synchrony, then the connection strength between them gets bigger. In our Bayesian-belief metaphor, the strength of a connection represents the brain's degree of belief that when one of them fires, so should the other one. Hebbian learning reinforces the brain's belief structure.

PSYCHOLOGISTS HAVE OBSERVED THAT WHEN: a person is told some new information, they don't just file it away in memory. That would be a disaster in evolutionary terms, because it's not a good idea to believe everything you're told. People tell lies, and they try to mislead others, often as part of a process aimed at bringing the others under their control. Nature tells lies too: that waving leopard tail may, on closer analysis, turn out to be a dangling vine or a fruit; stick insects pretend to be sticks. So when we receive new information, we assess it against our existing beliefs. If we're smart, we also assess the credibility of the information. If it comes from a trusted source, we're more likely to believe it; if not, less likely. Whether we accept the new information, and modify our beliefs accordingly, is the outcome of an internal struggle between what we already believe, how the new information relates to what we already believe, and how much confidence we have that the new information is true. Often this struggle is subconscious, but we can also reason consciously about the information.

In a bottom-up description, what's happening is that complex arrays of neurons are all firing and sending signals to each other. How those signals cancel each other out, or reinforce each other, determines whether the new information sticks, and the connection strengths change to accommodate it. This already explains why it's very hard to convince 'true believers' that they're wrong, even when the evidence seems overwhelming to everyone else. If someone has a strong belief in UFOs and the United States government puts out a press release explaining that an apparent sighting was actually a balloon experiment, the believer's Bayesian brain will almost certainly discount the explanation as propaganda. The press release will very possibly reinforce their belief that they don't trust the government on this issue, and they'll congratulate themselves on not being so gullible as to believe government lies. Beliefs cut both ways, so someone who doesn't believe in UFOs will accept the explanation as fact, often without independent verification, and the information will reinforce their belief that they don't trust UFO nuts. They'll congratulate themselves on not being so gullible as to believe in UFOs.

Human culture and language have made it possible for the belief systems of one brain to be transferred into another. The process is neither perfectly accurate nor reliable, but it's effective. Depending on the beliefs concerned and whoever is analysing the process, it goes by many names: education, brainwashing, bringing up the kids to be good people, the One True Religion. The brains of young children are malleable, and their ability to assess evidence is still developing: consider Santa Claus, the Tooth Fairy, and the Easter Bunny – though children are quite shrewd and many understand they have to play the game to get the reward. The Jesuit maxim 'Give me a child until he is seven and I will give you the man' has two possible meanings. One is that what you learn when you're young lasts longest; the other is that brainwashing innocent children to accept a belief system fixes it in their minds throughout adult life. Both are likely true, and from some viewpoints they're identical.

THE BAYESIAN BRAIN THEORY EMERGED from a variety of scientific areas: Bayesian statistics, obviously, but also machine intelligence and psychology. In the 1860s Hermann Helmholtz, a pioneer in the physics and psychology of human perception, suggested that the brain organises its perceptions by building probabilistic models of the external world. In 1983 Geoffrey Hinton, working in artificial intelligence, proposed that the human brain is a machine that makes decisions about the uncertainties it encounters when observing the outside world. In the 1990s this idea was turned into mathematical models based on probability theory, embodied in the notion of a Helmholtz machine. This isn't a mechanical device, but a mathematical abstraction, comprising two related networks of mathematically modelled 'neurons'. One, the recognition network, works from the bottom up; it's trained on real data and represents them in terms of a set of hidden variables. The other, a top-down 'generative' network, generates values of those hidden variables and hence of the data. The training process uses a learning algorithm to modify the structure of the two networks so that they classify the data accurately. The two networks are modified alternately, a procedure known as a wake–sleep algorithm.

Similar structures with many more layers, referred to as 'deep learning', are currently achieving considerable success in artificial intelligence. Applications include the recognition of natural speech by a computer, and computer victories in the oriental board game Go. Computers had previously been used to prove that the board game draughts is always a draw with perfect play. IBM's Deep Blue beat chess grandmaster and world champion Garry Kasparov in 1996, but lost the six-game series 4–2. After a major upgrade, it won the next series $3\frac{1}{2} - 2\frac{1}{2}$. However, those programs used brute force algorithms rather than the artificial intelligence algorithms used to win at Go.

Go is an apparently simple game with endless depths of subtlety, invented in China over 2500 years ago, and played on a 19×19 grid. One player has white stones, the other black; they place their stones in turn, capturing any they surround. Whoever surrounds the most territory wins. Rigorous mathematical analysis is very limited. An algorithm devised by David Benson can determine when a chain of stones can't be captured no matter what the other player does.[66] Elwyn Berlekamp and David Wolfe have analysed the intricate mathematics of endgames, when much of the board is captured and the range of available moves is even more bewildering than usual.[67] At that stage, the game has effectively split into a number of regions that scarcely interact with each other, and players have to decide which region to play in next. Their mathematical techniques associate a number, or a more esoteric structure, with each position, and provide rules for how to win by combining these values.

In 2015 the Google company DeepMind tested a Go-playing algorithm AlphaGo, based on two deep-learning networks: a value network that decides how favourable a board position is, and a policy network that chooses the next move. These networks were trained using a combination of games played by human experts, and games where the algorithm played against itself.[68] AlphaGo then pitted its electronic wits against Lee Sedol, a top professional Go player, and beat him, four games to one. The programmers found out why it had lost one game and corrected its strategy. In 2017 AlphaGo beat Ke Jie, ranked number one in the world, in a three-game match. One interesting feature, showing that deep-learning algorithms need not function like a human brain, was AlphaGo's style of play. It often did things that no human player would have considered – and won. Ke Jie remarked: 'After humanity spent thousands of years improving our tactics, computers tell us that humans are completely wrong... I would go as far as to say not a single human has touched the edge of the truth of Go.'

There's no logical reason why artificial intelligence should work in the same way as human intelligence: one reason for the adjective 'artificial'. However, these mathematical structures, embodied in electronic circuits, bear some similarity to the cognitive models of the brain developed by neuroscientists. So a creative feedback loop between artificial intelligence and cognitive science has emerged, each borrowing ideas from the other. And it's starting to look as though our brains and artificial ones do in fact work, sometimes and to some extent, by implementing similar structural principles. Down at the level of the materials they're made from, and how their signalling processes function, they're of course very different.

TO ILLUSTRATE THESE IDEAS IN A concrete setting, though with a more dynamic mathematical structure, consider visual illusions. Visual perception involves puzzling phenomena when ambiguous or incomplete information is presented to one or both eyes. Ambiguity is one type of uncertainty: we're not sure exactly what we're seeing. I'll take a quick look at two distinct types.

The first type was discovered by Giambattista della Porta in 1593, and included in his *De refractione* (On Refraction), a treatise on optics. della Porta put one book in front of one eye, and another in front of the other. He reported that he could read from one book at a time, and

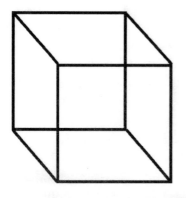 

Left: Necker cube. *Right*: Jastrow's rabbit/duck.

he could change from one to the other by withdrawing the 'visual virtue' from one eye and moving it to the other. This effect is now called binocular rivalry. It occurs when two different images, presented one to each eye, lead to alternating percepts – what the brain believes it's seeing – possibly of neither image separately.

The second type is illusions, or multistable figures. These occur when a single image, either static or moving, can be perceived in several ways. Standard examples are the Necker cube, introduced by the Swiss crystallographer Louis Necker in 1832, which appears to flip between two different orientations, and the rabbit/duck illusion invented by the American psychologist Joseph Jastrow in 1900, which flips between a not very convincing rabbit and an only slightly more convincing duck.[69]

A simple model of the perception of the Necker cube is a network with just two nodes. These represent neurons, or small networks of neurons, but the model is intended only for schematic purposes. One node corresponds to (and is assumed to have been trained to respond to) one perceived orientation of the cube, the other to the opposite orientation. These two nodes are linked to each other by inhibitory connections. This 'winner-takes-all' structure is important, because the inhibitory connections ensure that if one node is active, the other isn't. So the network comes to an unambiguous decision at any given time. Another modelling assumption is that this decision is determined by whichever node is most active.

Initially both nodes are inactive. Then, when the eyes are shown the

Necker cube image, the nodes receive inputs that trigger activity. However, the winner-takes-all structure means that both nodes can't be active at the same time. In the mathematical model they take turns: first one is more active, then the other. Theoretically these alternatives repeat at regular intervals, which isn't quite what's observed. Subjects report similar changes of percept, but they occur at irregular intervals. The fluctuations are usually explained as random influences coming from the rest of the brain, but this is open to debate.

The same network also models binocular rivalry. Now the two nodes correspond to the two images shown to the subject: one to the left eye, the other to the right eye. People don't perceive the two images superimposed on each other; instead, they alternate between seeing one of them and seeing the other. Again, this is what happens in the model, though with more regular timing of the switches between percepts.

If the mathematical model predicted only a switch between the two known possibilities, it wouldn't be terribly interesting. But in slightly more complicated circumstances, analogous networks behave in more surprising ways. A classic example is the monkey/text experiment of Ilona Kovács and colleagues.[70] A picture of a monkey (it looks suspiciously like a young orangutan, which is an ape, but everyone calls it a monkey) is cut into six pieces. A picture of blue text on a green background is cut into six similarly shaped pieces. Then three of the pieces in each image are swapped with the corresponding pieces in the other one to create two mixed images. These are then shown separately to the subject's left and right eyes.

What do they see? Most report seeing the two mixed images, alternating. This makes good sense: it's what happened with Porta's two books. It's as if one eye wins, then the other, and so on. But some subjects report alternation between a complete monkey image and complete text. There's a handwaving explanation of that: their brain 'knows' what a complete monkey and complete text should look like, so it fits suitable pieces together. But since it's seeing both mixtures, it still can't decide which one it's looking at, so it flips between them. However, this isn't very satisfactory, and it doesn't really explain why some subjects see one pair of images and others see a different pair.

A mathematical model sheds more light. It's based on a network model for high-level decision-making in the brain, proposed by the neuroscientist Hugh Wilson. I'll call models of this type Wilson

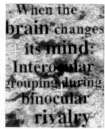

If the first two 'mixed' images are shown to each eye separately, some subjects see alternation between the last two complete images.

networks. In its simplest form, an (untrained) Wilson network is a rectangular array of nodes. These can be thought of as model neurons, or populations of neurons, but for modelling purposes they need not be given any specific physiological interpretation. In the rivalry setting, each column of the array corresponds to an 'attribute' of the image presented to the eye: a feature such as colour or orientation. Each attribute has a range of alternatives: for example, colour could be red, blue, or green; orientation could be vertical, horizontal, or diagonal. These discrete possibilities are the 'levels' of that attribute. Each level corresponds to a node in that attribute column.

Any specific image can be viewed as a combination of choices of levels, one for each relevant attribute. A red horizontal image combines the 'red' level of the colour column with the 'horizontal' level of the orientation column, for instance. The architecture of a Wilson network is designed to detect patterns by responding more strongly to 'learned' combinations of specific levels, one for each attribute. In each column, all pairs of distinct nodes are connected to each other by inhibitory couplings. Without further input or modification, this structure creates a winner-takes-all dynamic in the column, so that usually only one node is dynamically the most active. The column then detects the corresponding level of its attribute. Training by the images presented to the eyes is modelled by adding excitatory connections between the nodes corresponding to the appropriate combination of levels. In a rivalry model, such connections are added for both images.

Casey Diekman and Martin Golubitsky have shown that a Wilson network model of rivalry can sometimes have unexpected implications.[71] For the monkey/text experiment, the dynamics of the

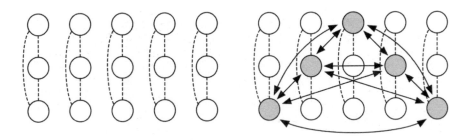

Left: Untrained Wilson network with five attributes, each having three levels. Dashed lines are inhibitory connections. *Right*: A pattern (shaded level for each attribute) is represented by excitatory connections (solid arrows) between those nodes. Adding these connections to the original network trains it to recognise that pattern.

network predicts that it can oscillate in two distinct ways. As we'd expect, it can alternate between the two learned patterns – the mixed images shown to the eye. But it can also alternate between a complete monkey and complete text. Which pair occurs depends on the connection strengths, suggesting that the difference between subjects is related to how strongly or weakly the corresponding populations of neurons are connected in the subject's brain. It's striking that the simplest Wilson network representing the experiment predicts exactly what's observed in experiments.

WILSON NETWORKS ARE SCHEMATIC MATHEMATICAL models, intended to shed light on how simple dynamic networks can in principle make decisions based on information received from the outside world. More strongly, some regions in the brain have a very similar structure to a Wilson network, and they seem to make their decisions in much the same way. The visual cortex, which processes signals from the eyes to decide what we're looking at, is a case in point.

Human vision doesn't work like a camera, whatever the school textbooks say. To be fair, the way the *eye* detects images is rather like a camera, with a lens that focuses incoming light on to the retina at the back. The retina is more like a charge-coupled device in a modern digital camera than an old-fashioned film. It has a large number of discrete receptors, called rods and cones: special light-sensitive neurons that respond to incoming light. There are three types of cone, and each is more sensitive to light in a specific range of wavelengths, that is, light of (roughly) a specific colour. The colours, in general terms, are red, green, and blue. Rods respond to low light levels. They respond most strongly to light around the 'light blue' or cyan wavelengths, but our visual system interprets these signals as shades of grey, which is why we don't see much colour at night.

Where human vision starts to differ significantly from a camera is what happens next. These incoming signals are transmitted along the optic nerves to a region of the brain called the visual cortex. The cortex can be thought of as a series of thin layers of neurons, and its job is to process the patterns of signals received from the eyes so that other regions of the brain can identify what they're seeing. Each layer responds dynamically to the incoming signals, much as a Wilson network responds to a Necker cube or the monkey/text pair of images. Those responses are transmitted down to the next layer, whose structure causes it to respond to different features, and so on. Signals also pass from the deeper layers to the surface ones, affecting how they respond to the next batch of signals. Eventually somewhere along this cascade of signals, *something* decides 'it's granny', or whatever. It might perhaps be one specific neuron, often referred to as a grandmother cell, or it might do this in a more sophisticated way. We don't yet know. Once the brain has recognised granny, it can pull up other information from other regions, such as 'help her off with her coat', or 'she always likes a cup of tea when she arrives', or 'she's looking a bit worried today'.

Cameras, when hooked up to computers, are starting to perform some tasks like this too, such as using facial recognition algorithms to tag your photos with the names of the people in them. So although the visual system isn't like a camera, a camera is becoming more and more like the visual system.

Neuroscientists have studied the wiring diagram of the visual cortex in some detail, and new methods of detecting connections in the brain will doubtless lead to an explosion of more refined results. Using special dyes that are sensitive to electrical voltages, they've mapped out the general nature of connections in the top layer V1 of the visual cortex in animals. Roughly speaking, V1 detects segments of straight lines in whatever the eyes are looking at, and it also works out in which direction the lines point. This is important for finding the boundaries of objects. It turns out that V1 is structured much like a Wilson network that has been trained using straight lines in various orientations. Each column in the network corresponds to a 'hypercolumn' in V1, whose attribute is 'direction of a line seen at this location'. The levels of that attribute are a coarse-grained set of directions that the line might point in.

The really clever part is the analogue of the learned patterns in a Wilson network. In V1, these patterns are longer straight lines, crossing the visual fields of many hypercolumns. Suppose a single hypercolumn detects a short piece of line somewhere at an angle of 60°, so the neuron for this 'level' fires. It then sends excitatory signals to neurons in neighbouring hypercolumns, but only to those at the same 60° level. Moreover, these connections link only to those hypercolumns that lie along the continuation of that line segment in V1. It's not quite that precise, and there are other weaker connections, but the strongest ones are pretty close to what I've described. The architecture of V1 predisposes it to detect straight lines and the direction in which they point. If it sees a piece of such a line, it 'assumes' that line will continue, so it fills in gaps. But it doesn't do this slavishly. If sufficiently strong signals from other hypercolumns contradict this assumption, they win instead. At a corner, say, where two edges of an object meet, the directions conflict. Send that information to the next layer down and you now have a system to detect corners as well as lines. Eventually, somewhere along this cascade of data, your brain recognises granny.

A FORM OF UNCERTAINTY THAT most of us have experienced at some point is: 'Where am I?' Neuroscientists Edvard and May-Britt Moser and their students discovered in 2005 that rat brains have special neurons, known as grid cells, that model their location in space. Grid cells live in a part of the brain whose name is a bit of a mouthful: the dorsocaudal medial entorhinal cortex. It's a central processing unit for location and memory. Like the visual cortex, it has a layered structure, but the pattern of firing differs from layer to layer.

The scientists placed electrodes in the brains of rats and then let them run around freely in an open space. When a rat was moving, they monitored which cells in its brain fired. It turned out that particular cells fire whenever the rat is in one of a number of tiny patches of the space ('firing fields'). These patches form a hexagonal grid. The researchers deduced that these nerve cells constitute a mental representation of space, a cognitive map that provides a kind of coordinate system, telling the rat's brain where the animal is. The activity of grid cells is updated continuously as the animal moves. Some cells fire whichever direction the rat is heading in; others depend on the direction and are therefore responsive to it.

We don't yet understand exactly how grid cells tell the rat where it is. Curiously, the geometric arrangement of the grid cells in its brain is irregular. Somehow, these layers of grid cells 'compute' the rat's location by integrating tiny movements as it wanders around. Mathematically, this process can be realised using vector calculations, in which the position of a moving object is determined by adding together lots of small changes, each with its own magnitude and direction. It's basically how sailors navigated by 'dead reckoning' before better navigational instruments were devised.

We know the network of grid cells can function without any visual input, because the firing patterns remain unchanged even in total darkness. However, it also responds quite strongly to any visual input. For example, suppose the rat runs inside a cylindrical enclosure, and there's a card on the wall to act as a reference point. Choose a particular grid neuron, and measure its grid of spatial patches. Then rotate the cylinder and repeat: the grid rotates through the same amount. The grids and their spacings don't change when the rat is placed in a new environment. However the grid cells compute location, the system is very robust.

In 2018 Andrea Banino and coworkers reported using deep-learning networks to carry out a similar navigational task. Their network had lots of feedback loops, because navigation seems to depend on using the output of one processing step as the input for the next – in effect a discrete dynamical system with the network as the function being iterated. They trained the network using recorded patterns of the paths that various rodents (such as rats and mice) had used when foraging, and provided it with the kind of information that the rest of the brain might send to grid neurons.

The network learned to navigate effectively in a variety of environments, and it could be transferred to a new environment without losing performance. The team tested its abilities by giving it specific goals, and in a more advanced setting by running it (in simulation, since the entire set-up is inside a computer) through mazes. They assessed its performance for statistical significance using Bayesian methods, fitting the data to mixtures of three distinct normal distributions.

One remarkable result was that as the learning process progressed, one of the middle layers of the deep-learning network developed similar patterns of activity to those observed in grid neurons, becoming active when the animal was in some member of a grid of spatial patches. Detailed mathematical analysis of the structure of the network suggested that it was simulating vector calculations. There's no reason to suppose it was doing that the way a mathematician would, writing down the vectors and adding them together. Nonetheless, their results support the theory that grid cells are critical for vector-based navigation.

MORE GENERALLY, THE CIRCUITS THAT the brain uses to understand the outside world are to some extent modelled on the outside world. The structure of the brain has evolved over hundreds of thousands of years, 'wiring in' information about our environment. It also changes over much shorter periods as we learn. Learning 'fine tunes' the wired-in structures. What we learn is conditioned by what we're taught. So if we're taught certain beliefs from a very early age, they tend to become wired into our brains. This can be seen as a neuroscientific verification of the Jesuit maxim mentioned earlier.

Our cultural beliefs, then, are strongly conditioned by the culture we grow up in. We identify our place in the world and our relations to those around us by which hymns we know, which football teams we support, what music we play. Any 'beliefs' encoded in our brain wiring that are common to most people, or which can be debated rationally in terms of evidence, are less contentious. But the beliefs we hold that don't have such support can be problematic unless we recognise the difference. Unfortunately those beliefs play an important role in our culture, which is one reason they exist at all. Beliefs based on faith, not evidence, are very effective for distinguishing Us from Them. Yes we all 'believe' that $2 + 2 = 4$, so that doesn't make me any different from you. But do you pray to the cat-goddess every Wednesday? I thought not. You're not One of Us.

This worked quite well when we lived in small groups, because nearly everyone we met did pray to the cat-goddess, and it was a good idea to be warned if not. But even when extended to tribes it became a source of friction, often leading to violence. In today's connected world, it's becoming a major disaster.

Today's populist politics has given us a new phrase for what used to be called 'lies' or 'propaganda'. Namely, Fake News. It's getting increasingly difficult to distinguish real news from fake. Vast computing power is placed in the hands of anyone with a few hundred dollars to spare. The widespread availability of sophisticated software democratises the planet, which in principle is good, but it often compounds the problem of differentiating truth from lies.

Because users can tailor what kind of information they see so that it reinforces their own preferences, it's increasingly easy to live in an information bubble, where the only news you get is what you want to be told. China Miéville parodied this tendency by taking it to extremes in *The City & the City*, a science-fiction/crime mash-up in which Inspector Borlú, a detective in the Extreme Crime Squad of the city of Besźel, is investigating a murder. He makes repeated visits to its twin city of Ul Qoma to work with the police there, crossing the border between them. At first, the picture you get is rather like Berlin before the wall came down, divided into East and West, but you slowly come to realise that the two halves of the city occupy *the same geographical space*. Citizens of each are trained from birth not to notice the other, even as they walk among its buildings and people. Today, many of us are doing the same on the internet, wallowing in confirmation bias, so that all of the information we receive reinforces the view that we're right.

Why are we so easily manipulated by fake news? It's that age-old Bayesian brain, running on embodied beliefs. Our beliefs aren't like files in a computer, which can be deleted or replaced at the twitch of a mouse. They're more like the hardware, wired in. Changing wired-in patterns is hard. The more strongly we believe, or even just *want* to believe, the harder it gets. Each item of fake news that we believe, because it suits us to, reinforces the strength of those wired-in connections. Each item that we don't want to believe is ignored.

I don't know a good way to prevent this. Education? What happens if a child goes to a special school that promotes a particular set of beliefs? What happens when it's forbidden to teach subjects whose factual status is clear, but which contradict beliefs? Science is the best route humanity has yet devised for sorting fact from fiction, but what happens if a government decides to deal with inconvenient facts by cutting funding for research on them? Federal funding for research on the effects of gun ownership is already illegal in the USA, and the Trump administration is considering doing the same for climate change.

It won't go away, guys.

One suggestion is that we need new gatekeepers. But an atheist's trusted website is anathema to a true believer, and vice versa. What happens if an evil corporation gets control of a website we trust? As always, this isn't a new problem. As the Roman poet Juvenal wrote in his *Satires* around 100 AD, *Quis custodiet ipsos custodes?* Who will guard the guards themselves? But the problem today is worse, because a single tweet can span the entire planet.

Perhaps I'm too pessimistic. On the whole, better education makes people more rational. But our Bayesian brains, whose quick-and-dirty survival algorithms served us so well when we lived in caves and trees, may no longer be fit for purpose in the misinformation age.

15

QUANTUM UNCERTAINTY

It is impossible to determine accurately both the position and the direction and
speed of a particle at the same instant.
Werner Heisenberg, *Die Physik der Atomkerne*

IN MOST AREAS OF HUMAN activity, uncertainty arises from ignorance.
Knowledge can resolve uncertainty, at least in principle. There are
practical obstacles: to predict the result of a democratic vote, we might
need to know what's going on in every voter's mind. But if we did
know that, we could work out who was going to vote, and what their
vote would be.

In one area of physics, however, the overwhelming consensus is that
uncertainty is an inherent feature of nature. No amount of extra
knowledge can make events predictable, because the system itself
doesn't 'know' what it's going to do. It just does it. This area is
quantum mechanics. It's about 120 years old, and it has totally
revolutionised not just science, but how we think about the relation
between science and the real world. It has even led some
philosophically minded people to question in what sense the real
world exists. Newton's greatest advance was to show that nature obeys
mathematical rules. Quantum theory shows us that even the rules can
be inherently uncertain. Or so almost all physicists assert, and they've
got plenty of evidence to support the claim. However, there are a few
chinks in the probabilistic armour. I doubt that quantum uncertainty
can ever be made predictable, but it might just have a deterministic
explanation. Before we look into these more speculative ideas in
Chapter 16, we need to sort out the orthodox story.

IT ALL BEGAN WITH A light bulb. Not a metaphorical one, hovering over some genius's head as inspiration struck: a real one. In 1894 several electric companies asked a German physicist, Max Planck, to develop the most efficient light bulb possible. Naturally, Planck started from basic physics. Light is a form of electromagnetic radiation, at wavelengths that the human eye can detect. Physicists knew the most efficient radiator of electromagnetic energy is a 'black body', characterised by a complementary property: it *absorbs* radiation of all wavelengths perfectly. In 1859 Gustav Kirchhoff had asked how the intensity of black-body radiation depends on the frequency of the emitted radiation and the temperature of the black body. Experimentalists made measurements, theorists devised explanations; the results disagreed. It was a bit of a mess, and Planck decided to tidy up.

His first attempt worked, but it didn't satisfy him because its assumptions were rather arbitrary. A month later he found a better way to justify them. It was a radical idea: electromagnetic energy isn't a continuous quantity, but a discrete one. It always comes in whole number multiples of a fixed, very tiny, amount. More precisely, for a given frequency, the energy is always a whole number times the frequency times a very small constant, now called Planck's constant and denoted by the symbol h. Its official value is 6.626×10^{-34} joule-seconds, that is, $0.0...0626$ with 33 zeros after the decimal point. One joule of energy increases the temperature of a teaspoon of water by about a quarter of a degree Celsius. So h is a very small quantity of energy indeed, so small that experimental energy levels still look continuous. Nonetheless, replacing a continuous range of energies by a discrete set of very finely spaced energies avoided a mathematical problem that had been giving the wrong results.

Planck didn't realise it at the time, but his curious assumption about energy was going to start a major revolution in the whole of science: quantum mechanics. A 'quantum' is a very small but discrete quantity. Quantum mechanics is the best theory we have of how matter behaves on very small scales. Although quantum theory fits experiments with amazing accuracy, much of what we know about the quantum world is distinctly puzzling. The great physicist Richard Feynman is reported to have said: 'If you think you understand quantum mechanics, you don't understand quantum mechanics.'[72]

For instance: the most obvious interpretation of Planck's formula is that light is made from tiny particles, now called photons, and the energy of a photon is its frequency multiplied by Planck's constant. Light comes in integer multiple of that value because you have to have a whole number of photons. This explanation makes excellent sense, but it raises a different question: How can a particle have a frequency? Frequency makes sense for waves. So is a photon a wave or a particle?

It's both.

GALILEO CLAIMED THAT NATURE'S LAWS are written in the language of mathematics, and Newton's *Principia* vindicated that opinion. Within a few decades, the mathematicians of continental Europe were extending this insight into heat, light, sound, elasticity, vibrations, electricity, magnetism, and fluid flow. The era of classical mechanics created by this explosion of mathematical equations contributed two major ingredients to physics. One was that of a particle – a piece of matter so small that for modelling purposes it can be thought of as a point. The other iconic concept was that of a wave. Think of a water wave travelling across the ocean. If there's not much wind, and it's far from land, the wave travels at a steady speed without changing its shape. The actual molecules of water that make up the wave don't move with it. They stay pretty much where they are. As the wave passes, water molecules move up and down and from side to side. They transmit that motion to nearby molecules, which move in a similar way, creating the same basic shape. So the *wave* travels, but the water doesn't.

Waves are everywhere. Sound is a pressure wave in the air. Earthquakes create waves in the ground, making buildings fall down. Radio signals, which give us television, radar, mobile phones, and the internet, are waves of electricity and magnetism.

So, it turned out, is light.

Towards the end of the 17th century, the nature of light was a matter of some controversy. Newton believed that light is made from lots of tiny particles. The Dutch physicist Christiaan Huygens offered strong evidence that light is a wave. Newton countered with ingenious particle explanations, and for about a century his views prevailed. Then it turned out that Huygens had been right all along. What eventually clinched the argument in favour of the wave camp was the phenomenon of interference. If light passes through a lens, or past the edge of a slit, it forms patterns: roughly parallel stripes of light and dark regions. It's easier to see this with a microscope, and it works better if the light all has a single colour.

Wave theory explains such phenomena in a simple, natural way: they're interference patterns. When two sets of waves overlap, their peaks reinforce each other, and so do their troughs, but a peak and a trough cancel out. You can see this easily by tossing two stones into a pond. Each stone creates a series of circular ripples, which spread outwards from the point of impact. Where these ripples cross, you get a complicated pattern more like a curved chequerboard, as in the picture.

All this seemed pretty convincing, and scientists accepted that light is a wave, not a particle. It was obvious. Then along came Planck, and suddenly it wasn't obvious any more.

THE CLASSIC PROOF THAT PHOTONS have a dual nature – sometimes particle, sometimes wave – emerged from a series of experiments. In 1801 Thomas Young imagined passing a beam of light through two thin parallel slits. If light is a wave, it will 'diffract' when it passes through a thin slit. That is, it will spread out on the far side, like those circular ripples on a pond. With two slits, diffraction should produce a characteristic interference pattern, like the one formed by dropping two stones close together.

Young's picture (overleaf) shows peaks as dark regions, troughs as white ones. Two concentric rings of waves, originating at slits A and B, overlap and interfere, leading to lines of peaks radiating towards C, D, E, and F. An observation at the right-hand edge of the picture would detect alternating bands of light and dark. Young didn't actually perform this experiment, but he demonstrated a similar one using a thin beam of sunlight split in half by a piece of card. The diffraction bands duly appeared. Young declared light to be a wave, and estimated the wavelengths of red and violet light from the sizes of the bands.

So far, this experiment just confirms that light is a wave. The next development was a slow burner, whose implications took a while to sink in. In 1909 Geoffrey Ingram Taylor, then an undergraduate,

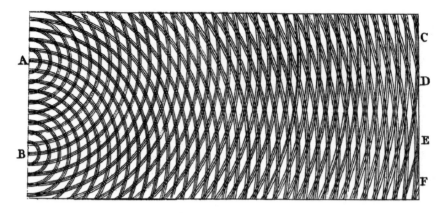

Young's drawing of two-slit interference based on observations of water waves.

performed a version of the double-slit experiment using a very weak light source diffracted on either side of a sewing needle. As with Young's card, the 'slits' were the regions on either side. Over a period of three months, a diffraction pattern built up on a photographic plate. His write-up doesn't mention photons, but the light was so weak that most of the time only a single photon passed by the needle, so the experiment was later interpreted as a proof that the pattern isn't caused by two photons interfering with each other. If so, it proves that a single photon can behave like a wave. Later still, Feynman argued that if you placed detectors to observe which slit the photon went through, the pattern should disappear. This was a 'thought experiment', not actually carried out. But putting it all together, it seemed that photons sometimes behaved like particles and sometimes like waves.

For a time, some quantum theory texts motivated the dual wave/particle nature of the photon by presenting the double-slit experiment and Feynman's afterthought as fact, even though neither experiment had actually been performed. In modern times they've been done properly, and the photons do what the textbooks said. The same goes for electrons, atoms, and (the current record) an 810-atom molecule. In 1965 Feynman wrote[73] that this phenomenon 'is impossible to explain in any classical way, and which has in it the heart of quantum mechanics'.

Many similar examples of quantum weirdness have been

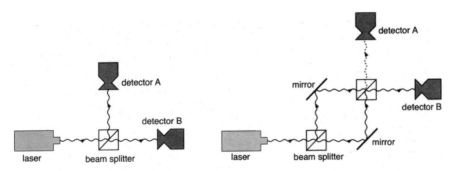

Left: Light is a particle. *Right:* Light is a wave.

discovered, and I'll summarise a pair of experiments that put the wave/particle problem in stark relief, from a paper by Roger Penrose.[74] They also illustrate some common observational techniques and modelling assumptions, which will be useful later. The key apparatus here is a favourite of experimentalists: a beam-splitter, which reflects half of the light shone on it, turning it through a right angle, but lets the other half pass through. A half-silvered mirror, in which the reflective metallic coating is so thin that some light can pass through, is one realisation. Often a glass cube is cut diagonally into two prisms, glued together along a diagonal face. The thickness of the glue controls the ratio between transmitted and reflected light.

In the first experiment, a laser emits a photon, which hits a beam-splitter. It turns out that exactly one of the detectors A and B observes a photon. This is particle-like behaviour: the photon has either been reflected and detected at A, or transmitted and detected at B. (The 'split' in 'beam-splitter' refers to the *probability* that the photon is reflected or transmitted. The photon itself remains intact.) This experiment doesn't make sense if the photon is a wave.

The second experiment uses a Mach–Zehnder interferometer: two beam-splitters and two mirrors arranged in a square. If photons were particles, we'd expect half of the photons to be reflected at the first beam-splitter, and the other half to be transmitted. Then the mirrors would send them to the second beam-splitter, and they'd have a 50% chance of going to A and a 50% chance of going to B. However, that's not what's observed. Instead, B always registers a photon, but A never does. This time the behaviour makes perfect sense if a photon is a wave, which splits into two smaller waves at the first beam-splitter.

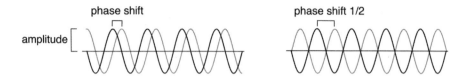

Left: Amplitude and relative phase shift of two waves. *Right*: A phase shift of 1/2 aligns peaks with troughs, so the waves cancel when superposed.

Each of them hits the second beam-splitter, to be split again. Detailed calculations, which I'll sketch in a moment, show that the two waves heading for detector A are out of phase (where one has a peak, the other has a trough) and cancel each other out. The two waves heading to B are in phase (their peaks coincide), and recombine to give a single wave – a photon.

So experiment one seems to prove that a photon is a particle but not a wave, and experiment two seems to prove that a photon is a wave but not a particle. You can see why physicists were baffled. Remarkably, they found a sensible way to fit it all together. Here's a quick informal sketch of the mathematics; it's not intended as a literal description of the physics. The wave function is expressed using complex numbers, of the form $a + ib$ where a and b are ordinary real numbers and i is the square root of minus one.[75] The main point to bear in mind is that when a quantum wave is reflected, either in a mirror or by a beam-splitter, its wave function is multiplied by i. (This follows, though not obviously, from the assumption that the beam-splitter is lossless: all photons are either transmitted or reflected.[76])

A wave has an amplitude, how 'high' it is, and a phase, which tells us where its peak is located. If we move the peak along a bit, this is a 'phase shift', and it's expressed as a fraction of the period of the wave. Waves whose phases differ by 1/2 cancel each other out; waves with the same phase reinforce each other. In wave terms, multiplying the wave function by i is like a phase shift of 1/4, because $i^4 = (-1)^2 = 1$. When the wave is transmitted, passing through the beam-splitter without being reflected, it's unchanged – phase shift 0.

At each successive transmission or reflection, the phase shifts add together. At beam-splitters, the wave becomes two half-waves, one going each way. The reflected wave shifts its phase by 1/4, while the transmitted one keeps the same phase. The diagram shows the

212

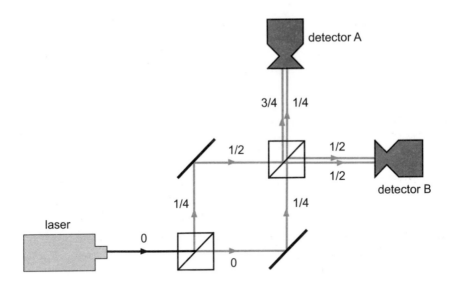

Paths through the apparatus. Half-waves shown in grey. The numbers are the phase shifts.

resulting paths through the apparatus. Half-waves are in grey, and the numbers are the phase shifts. Every reflection along a path adds another 1/4 to the total phase shift. Tracing the paths and counting reflections, you can check that detector A receives two half-waves, with phases 1/4 and 3/4. This is a phase difference of 1/2, so they cancel out and nothing is observed. Detector B also receives two half-waves, but now the phases are 1/2 and 1/2. This is a phase difference of 0, so they combine to give a single wave – and B observes a photon.

Magic!

The full calculations can be found in Penrose's paper. Similar methods apply to, and give results in agreement with, innumerable experiments – a remarkable success for the mathematical machinery. It's hardly surprising that physicists consider quantum theory to be a triumph of human ingenuity, as well as proof that nature, on its smallest scales, bears little resemblance to the classical mechanics of Newton and his successors.

HOW CAN A PARTICLE ALSO be a wave?

The usual answer from quantum theory is that the right kind of wave can behave like a particle. In fact, there's a sense in which a *single* wave is a slightly fuzzy particle. The peak travels without changing shape, which is exactly how a particle moves. Louis de Broglie and Erwin Schrödinger, two pioneers of quantum theory, represented a particle as a little bunch of waves, concentrated in small region, travelling along and wobbling up and down but staying together. They called this a wave packet. In 1925 Schrödinger devised a general equation for quantum waves, now named after him, publishing it the following year. It applies not just to subatomic particles, but to any quantum system. To find out what the system does, write down the appropriate version of Schrödinger's equation, and solve it to get the system's wave function.

In mathematical parlance, Schrödinger's equation is linear. That is, if you take a solution and multiply it by a constant, or add two solutions together, the result is also a solution. This construction is called *superposition*. Something similar happens in classical physics. There, although two particles can't be in the same place at the same time, two waves can happily coexist. In the simplest versions of the wave equation, solutions superpose to give solutions. As we've seen, one effect of superpositions is the creation of interference patterns. This property of Schrödinger's equation implies that its solutions are best thought of as waves, leading to the term 'wave function' for the quantum state of the system.

Quantum events occur on very small spatial scales, and they can't be observed directly. Instead, our knowledge of the quantum world is inferred from effects that we *can* observe. If it were possible to observe the entire wave function of, say, an electron, many quantum puzzles would vanish. However, this seems not to be possible. Certain special *aspects* of the wave function are open to observation, but not the whole thing. In fact, once you observe one aspect, the others are either unobservable, or have changed so much that the second observation bears no useful relation to the first.

These observable aspects of the wave function are called *eigenstates*. The name is a German/English hybrid, meaning something like 'characteristic states', and it has a precise mathematical definition. Any wave function can be built up by adding eigenstates together. Something similar happens in Fourier's heat equation, but it's easier to visualise for the closely related wave equation that models a vibrating violin string. The analogues of eigenstates are sine functions, like the picture on page 212, and any shape of wave can be built up by adding suitable combinations together. The basic sine wave is the fundamental pure note that the violin string produces; the other more closely spaced sine waves are its harmonics. In classical mechanics, we can measure the shape of the string in its entirety. But if you want to observe the state of a quantum system, you must first choose an eigenstate, and then measure just that component of the wave function. You can measure a different one afterwards, but the first observation disturbs the wave function, so the first eigenstate has probably changed by then. Although a quantum state can be (and usually is) a superposition of eigenstates, the outcome of a quantum measurement must be a pure eigenstate.

An electron, for example, has a property known as *spin*. The name comes from an early mechanical analogy and could have been totally different without any loss of comprehension – one reason why later quantum properties ended up with names like 'charm' and 'bottom'. Electron spin shares one mathematical property with classical spin: it has an axis. The Earth spins on its axis, giving us day and night, and the axis is tilted at an angle of 23·4° to the planet's orbit. An electron also has an axis, but it's a mathematical construct that can point in any direction at any time. The magnitude of the spin, on the other hand, is always the same: 1/2. At least, that's the associated 'quantum number', which is either a whole number or half a whole number for any quantum particle, and always the same for a given kind of particle.[77] The superposition principle means that the electron can spin *simultaneously* about many different axes – until you measure it. Choose an axis, measure the spin: you get either +1/2 or −1/2, because each axis points in two directions.[78]

That's weird. Theory says that the state is pretty much always a superposition; observations say that it's not. In some areas of human activity, this would be considered a massive discrepancy, but in quantum theory it's something you have to accept in order to get anywhere. And when you do, you get such beautiful results that you'd be crazy to reject the theory. Instead, you accept that the very act of measuring a quantum system somehow destroys features of the thing you're trying to measure.

One of the physicists grappling with such issues was Niels Bohr, who worked at the institute he founded in Copenhagen in 1921: the

Institute for Theoretical Physics, renamed the Niels Bohr Institute in 1993. Werner Heisenberg worked there in the 1920s, and in 1929 he gave a lecture in Chicago referring (in German) to the 'Copenhagen spirit of quantum theory'. This led, in the 1950s, to the term 'Copenhagen interpretation' of quantum observations. This maintained that when you observe a quantum system, the wave function is forced to collapse into a single-component eigenstate.

SCHRÖDINGER WASN'T TERRIBLY HAPPY ABOUT wave functions collapsing, because he believed they were real physical things. He came up with his famous thought experiment featuring a cat to argue against the Copenhagen interpretation. It involves yet another example of quantum uncertainty, radioactive decay. Electrons in an atom exist at specific energy levels. When they change level, the atom emits or absorbs energy, in the form of various particles, among them photons. In a radioactive atom, transitions of this kind can become violent enough to spit particles out of the nucleus and change the atom to a different element. This effect is called radioactive decay. That's why nuclear weapons and power stations work.

Decay is a random process, so the quantum state of a single radioactive atom, not currently being observed, is a superposition of 'not decayed' and 'decayed'. Classical systems don't behave like this. They exist in definite observable states. We live in a world which on human scales is (mostly) classical, but on sufficiently small scales is entirely quantum. How does this happen? Schrödinger's thought experiment played the quantum world off against the classical one. Put a radioactive atom (the quantum bit) inside a box, with a cat, a flask of poison gas, a particle detector, and a hammer (a classical system). When the atom decays, the detector trips the hammer, which smashes the flask, with sad consequences for puss.

If the box is impermeable to any method of observation, the atom is in a superposition of 'not decayed' and 'decayed'. Therefore, said Schrödinger, the cat must also be in a superposition: both 'alive' and 'dead' in suitable proportions.[79] Only when we open the box and observe what's inside do we collapse the wave function of the atom, hence also of the cat. Now it's either dead or alive, depending on what the atom did. Similarly, we find out that the atom either decayed or it didn't.

I don't want to go into all the ins and outs of this thought experiment here,[80] but Schrödinger didn't think a half-alive, half-dead cat made sense. His deeper point was that no one could explain *how* the wave function collapses, and no one could explain why large quantum systems, such as a cat considered as a huge collection of fundamental particles, seem to become classical. Physicists have performed experiments with ever larger quantum systems to show that superposition does occur. Not yet with cats, but they've done it with everything from electrons to very small diamond crystals. Simon Gröblacher is hoping to do it with tardigrades – tiny creatures also called 'water bears' or 'moss piglets', which are extraordinarily robust – by putting them on a quantum trampoline.[81] (Seriously. I'll come back to that.) However, these experiments don't answer Schrödinger's question.

The central philosophical issue is: What is an observation? Assuming the Schrödinger scenario, for the sake of argument, does the cat's wave function collapse as soon as the detector inside the box 'observes' the decay? Does it collapse when the *cat* notices the poison gas? (A cat can be an observer: one of our cats was an obsessive observer of goldfish.) Or does it wait until a human opens the box to see what's inside? We can argue for any of those, and if the box really is impermeable, there's no way to tell. Put a video camera inside? Ah, but you can't find out what it recorded until you open the box. Maybe it was in a superposition of 'recorded cat living' and 'recorded cat dying' until you observed it. Maybe its state had already collapsed as soon as the atom decayed. Maybe it collapsed somewhere in between.

The issue 'what is a quantum observation' is still unresolved. The way we model it in the mathematics is crisp and tidy, and bears no serious resemblance to how an actual observation is made, since it assumes the measuring apparatus isn't a quantum system. The way we ought to think about it is ignored by most physicists and hotly disputed by the rest. I'll come back to this debate in Chapter 16. For now, the main things to remember are the superposition principle, the fact that an eigenstate is all we can usually measure, and the unresolved nature of a quantum observation.

DESPITE THESE FOUNDATIONAL ISSUES, QUANTUM mechanics really took off. In the hands of a few brilliant pioneers, it explained a long list of experiments, which had previously been baffling, and motivated lots of new ones. Albert Einstein used quantum theory to explain the photoelectric effect, in which a beam of light hitting a suitable metal creates electricity. He got the Nobel Prize for this work. Ironically, he was never entirely happy about quantum theory. What worried him was uncertainty. Not in his mind: in the theory itself.

Mechanical quantities (classical or quantum) come in naturally associated pairs. For instance, position is associated with momentum (mass times velocity), and velocity is the rate of change of position. In classical mechanics you can measure both of these quantities simultaneously, and in principle those measurements can be as accurate as you wish. You just have to take care not to disturb the particle too much when you measure what it's doing. But in 1927 Heisenberg argued that quantum mechanics isn't like that. Instead, the more accurately you measure a particle's position, the less accurately you can determine its speed, and vice versa.

Heisenberg gave an informal explanation in terms of the 'observer effect': the act of observation disturbs what you're observing. It helped to convince people he was right, but actually it's an oversimplification. The observer effect arises in classical mechanics too. To observe the position of a football you can shine a light on it. The impacting light is reflected – and the ball slows down very slightly. When you subsequently measure the speed, say by timing how long it takes for the ball to travel one metre, it's a tiny bit lower than it was just before you shone the light. So measuring the ball's position alters the measurement of its speed. Heisenberg pointed out that in classical physics, careful measurements render this change negligible. But in the quantum realm, measurements are more like giving the football a hefty kick. Your foot now tells you where it *was*, but you have no idea where it's gone.

It's a neat analogy, but technically it's wrong. Heisenberg's limitation on the accuracy of quantum measurements runs much deeper. In fact, it occurs for any wave phenomenon, and it's further evidence for the wave nature of matter on very small scales. In the quantum world, it's stated formally as Heisenberg's uncertainty principle. It was formulated mathematically by Hesse Kennard in

1927, and by Hermann Weyl a year later. It says that uncertainty in position times uncertainty in momentum is at least $h/4\pi$, where h is Planck's constant. In symbols:

$$\sigma_x \sigma_p \geqslant h/4\pi$$

where the sigmas are standard deviations, x is position, and p is momentum.

This formula shows that quantum mechanics involves an inherent level of uncertainty. Science proposes theories and tests them with experiments. The experiments measure quantities predicted by the theory, to see if they're correct. But the uncertainty principle says that certain combinations of measurements are *impossible*. This isn't a limitation of current apparatus: it's a limitation of nature. So some aspects of quantum theory can't be verified experimentally.

To add to the weirdness, Heisenberg's principle applies to some pairs of variables, but not to others. It tells us that certain pairs of 'conjugate' or 'complementary' variables, such as position and momentum, are inextricably linked. If we measure one very accurately, the way they're linked mathematically implies that we can't also measure the other one very accurately. In some cases, however, two different variables can be measured simultaneously, even in a quantum world.

WE NOW KNOW THAT, CONTRARY to Heisenberg's explanation, the uncertainties expressed by the uncertainty principle don't come from the observer effect. In 2012 Yuji Hasegawa measured the spins of groups of neutrons, finding that the act of observation didn't create the amount of uncertainty prescribed by Heisenberg.[82] In the same year a team under Aephraim Steinberg managed to make measurements on photons that were so delicate, they introduced less uncertainty on individual photons than the uncertainty principle specifies.[83] However, the mathematics remains correct, because the total uncertainty about what the photons are doing still exceeds the Heisenberg limit.

The experiment doesn't use position and momentum, but a subtler property called polarisation. This is the direction in which the wave representing a photon is vibrating. It might be up and down, or side to side, or in some other direction. Polarisations in directions at right angles to each other are conjugate variables, so by the uncertainty principle you can't measure them simultaneously to arbitrarily high precision. The experimenters made a weak measurement of a photon's polarisation in one plane, which didn't disturb it much (like tickling a football with a feather). The result wasn't very accurate, but it gave some estimate of the direction of polarisation. Then they measured the same photon's polarisation in the second plane in the same way. Finally, they measured its polarisation in the original direction using a strong measurement (a hefty kick) that gave a very accurate result. That told them how much the weak measurements had disturbed each other. This final observation disturbed the photon a lot, but that didn't matter by then.

When these observations were repeated many times, the measurement of one polarisation didn't disturb the photon as much as Heisenberg's principle states. The actual disturbance could be half as great. This doesn't contradict the principle, however, because you can't measure *both* states sufficiently accurately. But it shows that it's not always the act of measurement that creates the uncertainty. It's there already.

COPENHAGEN ASIDE, SUPERPOSITIONS OF WAVE functions seemed straightforward until 1935, when Einstein, Boris Podolsky, and Nathan Rosen published a famous paper on what's now called the EPR paradox. They argued that according to the Copenhagen interpretation, a system of two particles must violate the uncertainty principle, unless a measurement made on one of them has an instantaneous effect on the other – no matter how widely separated they are. Einstein declared this to be 'spooky action at a distance', because it's not consistent with the basic relativistic principle that no signal can travel faster than light. Initially, he believed that the EPR paradox disproved the Copenhagen interpretation, so quantum mechanics was incomplete.

Today, quantum physicists see things very differently. The effect revealed by EPR is genuine. It arises in a very specific setting: two (or more) 'entangled' particles or other quantum systems. When particles are entangled, they lose their own identities, in the sense that any feasible observation refers to the state of the whole system, not to individual components. Mathematically, the state of the combined system is given by the 'tensor product' of the states of the components (I'll try to explain that in a moment). The corresponding wave function, as usual, gives the probability of observing the system in any given state. But the states themselves don't split up into observations of the separate components.

Tensor products work roughly like this. Suppose two people have a hat and a coat. Hats can be red or blue; coats can be green or yellow. Each person chooses one of each, so the 'state' of their attire is a pair such as (red hat, green coat) or (blue hat, yellow coat). In a quantum universe, states of hats can be superposed, so that '1/3 red + 2/3 blue hat' makes sense, and similarly for coats. The tensor product extends superpositions to the pairs (hat, coat). The mathematical rules tell us that for a fixed choice of coat colour, say green, the superposition of two hat states decomposes the entire system like this:

(1/3 red + 2/3 blue hat, green coat)

= 1/3(red hat, green coat) + 2/3(blue hat, green coat)

The same goes for a superposition of two coat states, for a fixed choice of hat. States like these in effect tell us that no significant interaction between hat and coat states is occurring. However, such interactions do occur for 'entangled' states such as

1/3(red hat, green coat) + 2/3(blue hat, yellow coat)

The rules of quantum mechanics predict that measuring the colour of the hat collapses not just the state of the hat, but that of the entire hat/coat system. This instantly implies constraints on the state of the coat.

The same goes for pairs of quantum particles: one for the hat, one for the coat. The colours are replaced by variables such as spin or polarisation. It might seem that somehow the particle that is measured *transmits* its state to the other one, affecting any measurement made on it. However, the effect happens however far apart the particles may be. According to relativity, signals can't travel faster than light, but in one experiment they would have to travel at 10,000 times light speed to explain the effect. For that reason, the effect of entanglements on observations is sometimes called quantum teleportation. It's considered to be a distinctive feature – perhaps *the* distinctive feature – that shows how different the quantum world is from classical physics.

LET ME RETURN TO EINSTEIN, worried about spooky action at a distance. Initially he favoured a different answer to the enigma of entangled states: a hidden-variable theory. An underlying deterministic explanation. Think about tossing a coin, as in Chapter 4. The probabilistic heads/tails state is explained by a more detailed mechanical model of the coin, with variables of position and spin rate. These are unrelated to the binary head/tail variable. That shows up when we 'observe' the state of the coin by interrupting its trajectory with a table, a hand, or the ground. The coin isn't mysteriously flickering between heads and tails in some random manner; it's doing something far more alien.

Suppose that every quantum particle has a hidden dynamic, which determines the result of an observation in a similar manner. Suppose further that when two particles are initially entangled, their hidden dynamics are synchronised. Thereafter, at any given moment, they're both in the same hidden state. This remains true if they're separated. If the outcome of a measurement isn't random, but prescribed by that internal state, measurements made on both particles at the same moment must agree. There's no need for a signal to pass between them.

It's a bit like two spies who meet, synchronise their watches, and split up. If at some moment one of them looks at the watch and it reads 6.34 p.m., she can predict that the other one's watch also reads 6.34 p.m. at that instant. They can both take action at the same prearranged moment without any signal passing between them. The hidden dynamic of a quantum particle could work the same way, acting like a watch. Of course the synchrony has to be very accurate, or the two particles will get out of step, but quantum states *are* very accurate. All electrons have the same mass to many decimal places, for instance.

It's a neat idea. It's very close to how entangled particles are generated in experiments.[84] It doesn't just show that in principle a deterministic hidden-variable theory can explain entanglement without spooky action at a distance: it comes close to a proof that such a theory must exist. But when physicists decided that no hidden-variable theory is possible, as we'll see in the next chapter, it went on the back burner.

16

DO DICE PLAY GOD?

Chaos preceded Cosmos, and it is into Chaos without form and void that we have plunged.

John Livingston Lowes, *The Road to Xanadu*

PHYSICISTS HAD COME TO RECOGNISE that matter on its smallest scales has a will of its own. It can decide to change – from particle to wave, from a radioactive atom of one element to a different element altogether – spontaneously. No outside agency is needed: it just *changes*. No rules. It wasn't exactly a matter of God not throwing dice, as Einstein grumbled. It was worse than that. Dice may be an icon of randomness, but we saw in Chapter 4 that they're actually deterministic. Bearing that in mind, Einstein's complaint should really have been that God *does* throw dice, whose hidden dynamic variables determine how they fall. The quantum view, to which Einstein objected, was that God doesn't throw dice, but He gets the same results as if He had. Or, more accurately still, the dice throw themselves, and the universe is what happens as a result. Basically, the quantum dice are playing God. But are they metaphorical dice, embodiments of genuine randomness, or are they deterministic dice, bouncing chaotically across the fabric of the cosmos?

In the creation myths of ancient Greece, 'Chaos' referred to a formless primordial state that came before the creation of the universe. It was the gap that appeared when Heaven and Earth were pulled apart, the void below the Earth on which Earth rests, and – in Hesiod's *Theogony* – the first primordial god. Chaos preceded Cosmos. In the development of modern physics, however, Cosmos preceded Chaos. Specifically, quantum theory was invented and developed half a century before the possibility of deterministic chaos was properly understood.

DO DICE PLAY GOD?

Quantum uncertainty was therefore assumed, from the start, to be purely random, built into the fabric of the universe.

By the time chaos theory became widely known, the paradigm that quantum uncertainty is inherently random, without any deeper structure to explain it, and that there's no *need* for a deeper structure to explain it, had become so entrenched that even to question it was taboo. But I can't help thinking that everything might have been different if mathematicians had come up with chaos theory – as a well-developed branch of mathematics, not just the odd example dug up by Poincaré – *before* physicists started wondering about quanta.

The issue is where quantum randomness comes from. The orthodox view is that it doesn't come from anything; it's just there. The problem is to explain why, in that case, quantum events have such regular statistics. Every radioactive isotope has a precise 'half-life', the time it takes for half the atoms in a large sample to decay. How does a radioactive atom know what its half-life is supposed to be? What tells it when to decay? It's all very well saying 'chance', but in every other context, chance either reflects ignorance of the mechanisms that create events, or mathematical deductions from knowledge of those mechanisms. In quantum mechanics, chance *is* the mechanism.

There's even a mathematical theorem saying that it has to be: Bell's theorem, which probably would have won John Bell the Nobel Prize in Physics. It's widely believed that he was nominated for that honour in 1990, but nominations are kept secret and Bell died of a stroke before the winners were announced. But, as with most pronouncements of fundamentalist physics, if you dig deeper, it's not as straightforward as everyone claims. It's often said that Bell's inequality rules out any hidden-variable theory for quantum mechanics, but this statement is too broad. It does indeed rule out certain *kinds* of hidden-variable explanation, but not all. The proof of the theorem involves a series of mathematical assumptions, not all of which are made explicit. More recent work suggests that certain types of chaotic dynamics might, in principle, offer a deterministic mechanism that underlies quantum uncertainty. At the moment these are hints, not a definitive theory, but they suggest that if chaos had been discovered before quantum theory, a deterministic view might have become the orthodox one.

EVEN FROM THE EARLIEST DAYS, a few physicists challenged the orthodox view of quantum indeterminacy. And in recent years, new maverick ideas have arisen, providing an alternative to the prevailing view that some doors are best not opened. Roger Penrose, one of the world's leading physicists, is one of the few who feel uncomfortable about the current attitude to quantum uncertainty. In 2011 he wrote: 'Quantum mechanics has to live with not only deep puzzles of interpretation, but ... a profound internal inconsistency, which is the reason that one might believe that there's something serious to be attended to about the theory.'[85]

Attempts to suggest alternatives to quantum orthodoxy are generally met with deep-seated suspicion by the bulk of the physics community. This is an understandable reaction to generations of crackpot attacks on fundamental physics, and to occasional forays by philosophers who seek verbal explanations of the riddles of the quantum, while castigating physicists for getting it all wrong. There's an easy way to avoid all these issues, and the temptation to use it must be strong. Quantum mechanics is weird. Even physicists say so. They delight in just how weird it is. Obviously, anyone who disagrees with them is some dyed-in-the-wool classical mechanist who lacks the imagination to accept that the world can possibly be *that* weird. 'Stop asking silly questions and get on with the calculations' became the dominant attitude.

However, there's always been a counterculture. Some people, among them the best and brightest of the world's physicists, have kept asking the silly questions anyway. Not out of a lack of imagination, but an excess of it. They wonder whether the quantum world is even *weirder* than the orthodox description. Profound discoveries about the silly questions have rocked the foundations of physics. Books have been written, papers published. Promising attempts to address the deeper aspects of quantum reality have appeared; some so effective that they agree with virtually everything that's currently known about quantum theory, while adding a new level of explanation. This very success has been used as a weapon against them. Since there's no way to distinguish the new theory from the existing one through conventional experiments, the argument goes, the new one is pointless and we should stick to the existing one. This strangely asymmetric argument has the obvious counter: by the same reasoning, the old theory is pointless and we should all switch to the new one. At that point, the opposition switches its Bayesian brain off, and goes back to its time-honoured ways.

And yet ... there are just *too many* loose ends in the conventional description of the quantum world. Phenomena that seem to make no sense. Assumptions that contradict themselves. Explanations that don't explain. And underneath it all is an uncomfortable truth, swept under the carpet with undue haste because it's deeply embarrassing: the 'shut up and calculate' brigade don't really understand it either. The truth is, even quantum physicists of that persuasion *don't* just calculate. Before doing the complicated sums, they set up quantum-mechanical equations to model the real world, and the choice of those equations goes beyond mere rule-based calculation. For example, they model a beam-splitter as a crisp yes/no mathematical entity, external to the equations, which either transmits a photon or reflects it – yet leaves its state miraculously unchanged, aside from a quarter-period phase shift in the reflected wave. There are no crisp entities of this kind. A real beam-splitter, viewed on the quantum level, is an enormously complicated system of subatomic particles. When a photon passes through one, it interacts with this entire system. Nothing crisp about that. Yet, amazingly, the crisp model appears to work. I don't think anyone really knows why. The sums would be impossible: a beam-splitter contains too many particles. Shut up and calculate.

Most quantum equations involve this kind of modelling assumption, importing into the mathematics sharply defined objects that don't exist in a fuzzy quantum world. The content of the equations is emphasised, to the neglect of the context – the 'boundary conditions' that must be specified before the equations can be set up and solved. Quantum physicists know how to use their bag of mathematical tricks; they're polished performers who can carry out astonishingly complex calculations and get answers correct to nine decimal places. But few ask why it works so well.

IS THE WAVE FUNCTION REAL? Or, since we can't observe it as a whole, is it just a mathematical abstraction? Does it actually exist, or is it a convenient fiction – a physicist's version, perhaps, of Quetelet's average man? The average man doesn't exist; we can't wander up to the right door, knock on it, and come face to face with him. Nevertheless, this fictional character encapsulates a lot of information about real men. Maybe the wave function is like that. No electron actually has one, but they all behave as if they did.

For a classical analogy, consider a coin. It has a probability distribution: $P(H) = P(T) = 1/2$. This exists in the standard mathematical sense: it's a well-defined mathematical object, and it governs pretty much everything that can be said about repeated coin tosses. But does the distribution *exist* as a real physical object? It's not marked on the coin. You can't measure it all at once. Whenever you toss the coin you get a definite result. Heads. Heads again. Tails this time. The coin behaves *as if* its probability distribution is real, but the only way to measure the distribution using the coin-toss 'instrument' is to toss the coin over and over again and count what happens. From this, you *infer* the distribution.

If the coin just sat on the table, somehow switching randomly between heads and tails, all of this would be very mysterious. How would the coin *know* how to make the odds fifty-fifty? Something has to tell it. So either the probability distribution is embodied in the reality of the coin, or something deeper is going on, which we're not seeing, and the distribution is a sign that a deeper truth is there to be found.

In this case, we know how it works. The coin doesn't just sit on the table flickering between the two states H and T that we can measure. It flips over and over in the air. While it's doing that, its state is neither heads nor tails. It's not even a superposition of heads and tails, fifty-fifty. It's totally different: a position and a rotation rate in space, not an up/down choice on a table. We 'observe' the head/tail choice by making the coin interact with a 'measuring instrument', the table. (Or a human hand, or whatever arrests its descent.) There's a hidden world of spin, to which the table is totally oblivious. Within that hidden world, the fate of the coin is decided: it ends up heads if the heads side of the coin is uppermost as it hits the table, at whichever angle it hits. Tails otherwise. All of the fine detail of the movement in space is obliterated by the act of observation: literally, smashed flat.

Could quantum uncertainty be like that? It's a plausible idea. It's how Einstein hoped to explain entanglement. Might the spinning electron have some sort of *internal* dynamical state, a hidden variable that's not observed directly, but which decides what value the electron's spin will be assigned when it interacts with the measuring instrument? If so, the electron is much like the coin, whose hidden variable is its dynamical state. The random observable is just its final rest state, 'measured' by its hitting a table. The same idea could explain how a radioactive atom decays, randomly but with regular statistical patterns. It would be easy to invent a suitable chaotic dynamic.

In classical mechanics, the existence of such hidden variables, with their secret dynamic, explains how the coin knows to give heads half the time. That information is a mathematical consequence of the dynamic: the probability that the system ends up given dynamical state. (This is like an invariant measure but technically different: it's about the distribution of initial conditions in state space that lead to a specified observation.) When the coin is spinning, we can trot ahead mathematically into its entirely deterministic future, and work out whether it will hit the table as heads or tails. We then conceptually label the current state with that outcome. To find the chance it lands heads, we calculate the fraction of state space (or some chosen region) whose points are labelled 'heads'. That's it.

Virtually every occurrence of randomness in a dynamical system can be explained in terms of a natural probability measure associated with deterministic but chaotic dynamics. So why not quantum systems? Unfortunately for seekers of hidden-variable explanations, there's what looks like a pretty good answer.

IN THE COPENHAGEN INTERPRETATION, IT'S considered pointless to speculate about hidden variables, on the grounds that any attempt to observe the 'inner workings' of atomic and subatomic processes must disturb them so much that the observations are meaningless. But this is hardly a knock-down argument. Today we routinely observe aspects of the inner structure of quantum particles using accelerators. They're costly – a cool 8 billion euros for the Large Hadron Collider, which discovered the Higgs boson – but not impossible. In the 18th century Auguste Comte argued that we can never find out the chemical composition of stars, but no one argued that stars don't *have* a chemical composition. It then turned out that Comte was spectacularly (and spectrographically) wrong: the chemical composition of stars is one of the main things we *can* observe. Spectral lines in the light from the star betray the chemical elements inside it.

The Copenhagen interpretation was heavily influenced by logical positivism, a view of the philosophy of science that prevailed in the early 20th century, which maintained that nothing can be considered to exist unless you can measure it. Scientists studying animal behaviour accepted this view, which led them to believe that anything an animal does is controlled by some mechanistic 'drive' in its brain. A dog doesn't drink water from a bowl because it's thirsty; it has a drive to drink, which switches on when its hydration level drops below a critical value. Logical positivism was itself a reaction to its opposite: anthropomorphism, a tendency to assume that animals have emotions and motivations just like humans. But it was an overreaction, transforming intelligent organisms into mindless machines. A more nuanced view is now current. For example, experiments carried out by Galit Shophat-Ophir suggests that male fruit flies experience pleasure during sex. 'The sexual reward system is very ancient machinery,' he says.[86]

Possibly the founders of quantum mechanics also overreacted. Within the last few years, scientists have found tricky ways to tease quantum systems into revealing rather more of their internal workings than was envisaged in Bohr's day. We encountered one in Chapter 15, Aephraim Steinberg's method for beating the uncertainty principle. It now seems accepted that the wave function is a real physical feature: very hard to observe in detail, perhaps impossibly so, but not merely a useful fiction. So rejecting hidden variables on the ground that a few famous physicists in Copenhagen in the 1920s decided that it can't be done is no more sensible than rejecting an internal chemistry for stars because Comte said you can never find out what it is.

There is, however, a better reason why most physicists reject hidden variables as the explanation of quantum uncertainty. Any such theory has to be consistent with everything currently known about the quantum world. Which is where John Bell's epic discovery of his inequality comes into the story.

IN 1964 BELL PUBLISHED WHAT is widely considered to be one of the most important papers on hidden-variable theories of quantum mechanics: 'On the Einstein Podolsky Rosen paradox'.[87] Bell was motivated by an earlier attempt by John von Neumann in 1932, whose book *Mathematische Grundlagen der Quantenmechanik* (Mathematical Foundations of Quantum Mechanics) contained a proof that no hidden-variable theory of quantum mechanics is possible.

The mathematician Grete Hermann spotted a flaw in the argument in 1935, but her work had no impact and the physics community accepted von Neumann's proof without question for decades.[88] Adam Becker wonders whether her gender was one reason, at a time when women were mostly forbidden to teach in universities.[89] Hermann had been a doctoral student of the greatest female mathematician of the period, Emmy Noether at Göttingen University. Noether started lecturing in 1916, nominally as a teaching assistant to David Hilbert, but she wasn't paid to give lectures until 1923. Anyway, Bell independently noticed that von Neumann's proof was incomplete. Attempting to find a hidden-variable theory, he instead discovered a much stronger impossibility proof. His central result rules out the possibility of any hidden-variable model of quantum indeterminacy that satisfies two basic conditions, both entirely reasonable in a classical context:

- *Reality*: microscopic objects have real properties determining the outcome of quantum measurements.
- *Locality*: reality in any given location is not affected by experiments performed simultaneously at a distant location.

Given these assumptions, Bell proved that certain measurements must be related by an inequality: a mathematical expression asserting that some combination of observable quantities is less than or equal to some other combination. It follows that if an experiment produces measurements that violate the inequality, then one of three things must occur: either the condition of reality fails, or locality fails, or the presumed hidden-variable theory doesn't exist. When experimental work produced results inconsistent with Bell's inequality, hidden-variable theories were declared dead. Quantum physicists reverted to 'shut up and calculate' mode, satisfied that the quantum world is so weird that this is all you can ever do, and anyone who wanted further explanations was wasting their own time and everybody else's.

I don't want to get too tied up in mathematical details, but we need to see roughly how Bell's proof proceeds. It's been revised and reworked many times, and these variations are collectively known as Bell inequalities. The current standard set-up involves the famous cryptographic pair Alice and Bob, who are making observations on pairs of entangled particles that have interacted and then separated. Alice measures one, Bob the other. For definiteness, assume they're measuring spins. The key ingredients are:

- A space of hidden variables. These represent a hypothetical internal mechanism within each particle, whose state determines the outcomes of measurements, but which is not directly observed. It's assumed that this space carries its own measure, telling us the probability that the hidden variables lie in some specified range. This set-up isn't deterministic, as stated, but we can include deterministic models if we specify a dynamic on the hidden variables and then use an invariant measure for the probabilities.

- Each of Alice and Bob has a detector, and chooses a 'setting'. Setting a is the axis about which Alice measures spin; setting b is the axis about which Bob measures spin.

- Observed correlations between the spins measured by Alice and those measured by Bob. These are quantities that quantify how often they both obtain the same result 'spin up' or 'spin down', compared to getting opposite results. (They're not quite the same as statistical correlation coefficients, but perform the same task.)

We can envisage three such correlations: the experimentally observed one, the one predicted by standard quantum theory, and the one predicted by some hypothetical hidden-variable theory. Bell considered three particular axes a, b, c, and calculated the hidden-variable predictions for their correlations in pairs, denoted $C(a,b)$ and so on. By relating these correlations to the presumed probability distribution on the space of hidden variables, he showed on general mathematical grounds that whatever the hidden-variable theory might be, the correlations must satisfy the inequality:[90]

$$C(a,\ c) - C(b,\ a) - C(b,\ c) \leqslant 1$$

This inequality is characteristic of the hidden-variable world. According to quantum theory, it's false for the quantum world. Experiments confirm that it's false in the real world. Score 1–0 for quantum v. hidden variables.

A classical analogy may help to explain why a condition of this general kind must apply. Suppose that three experimenters are tossing coins. The results are random, but either the coins or the devices tossing them have somehow been engineered so that they produce highly correlated results. Alice and Bob get the same result 95% of the time; Bob and Charlie get the same result 95% of the time. So Alice and Bob disagree on 5% of tosses, and Bob and Charlie disagree on 5% of tosses. Therefore Alice and Charlie can disagree on at most $5 + 5 = 10\%$ of tosses, so they must get the same result at least 90% of the time. This is an example of the Boole–Fréchet inequality. Bell's inequality follows roughly similar reasoning in a quantum context, but it's less straightforward.

Bell's inequality marked a vital stage in the investigation of hidden-variable theories, distinguishing their main features and ruling out theories that tried to do too much. It spoilt Einstein's simple explanation of the paradoxical features of entanglement: that the particles have synchronised hidden variables. Since there's no hidden dynamic, there can't be anything to synchronise.

Nevertheless, if it were possible to circumvent Bell's theorem, entanglement would make so much more sense. It seems a game worth playing. Let's look at some potential loopholes.

A WAVE IS HEADING TOWARDS a barrier that has two small slits, very close together. Different regions of the wave pass through each slit, emerging on the far side, and spread out. The two emerging waves overlap in a complex pattern of alternating peaks and troughs. This is a diffraction pattern, and it's what we'd expect of a wave.

A tiny particle is heading towards a barrier that has two small slits, very close together, so it either passes through one or through the other. Whichever slit it passes through, it can change direction in an apparently random way. But if observations of the position of the particle are averaged out over large numbers of repetitions, they too form a regular pattern. Bizarrely, it looks just like the diffraction pattern for a wave. This isn't at all what we'd expect of a particle. It's weird.

You'll recognise this as a description of one of the first experiments that revealed how weird the quantum world is: the famous double-slit experiment, which shows that a photon behaves like a particle in some circumstances, but like a wave in others.

It is. But it's *also* a description of a more recent experiment, which has nothing to do with quantum theory at all. If anything, that's even weirder.

In the experiment, the particle is a tiny droplet of oil, and the waves travel across a bath of the same oil. Amazingly, the droplet bounces on top of the waves. Ordinarily, if an oil droplet hits a bath of the same fluid, we expect it to merge into the bath and disappear. However, very small droplets of oil can be made to sit on top of a bath filled with the same oil without being swallowed up. The trick is to make the bath vibrate rapidly in the vertical direction, for example by placing it on top of a loudspeaker. There are forces in fluids that to some extent oppose mergers, and the main one here is surface tension. A fluid surface acts rather like a thin, soft elastic membrane; a deformable bag that holds the surface together. When a droplet touches the bath, surface tension tries to keep the two separate, while the force of gravity, among others, tries to make them merge. Which one wins depends on the circumstances. Vibrating the bath creates waves on the surface. When the droplet going down hits a wave coming up, the impact can overcome the tendency to merge, so the droplet bounces. With suitably small droplets and the right vibrational amplitudes and frequencies, the droplet bounces in resonance with the waves. This effect is very robust – at a frequency of 40 hertz, 40 times per second, the droplet usually bounces 20 times per second (half as fast, for mathematical reasons I won't go into). The droplet can remain intact for millions of bounces.

In 2005 Yves Couder's research group began to investigate the physics of bouncing droplets. The droplets are very small by comparison with the bath, and are observed using a microscope and a slow-motion camera. By changing the amplitude and frequency of the vibration, the droplet can be made to 'walk', travelling slowly in a straight line. This happens because it gets slightly out of sync with the wave, so instead of hitting it exactly at the peak, it bounces at a slight angle. The wave pattern moves as well, and if the numbers are right, exactly the same thing happens the next time they meet. Now the droplet is behaving like a moving particle (and, of course, the wave is behaving like a moving wave).

A group under John Bush took Couder's work further, and between them the two teams made some very curious discoveries. In particular, bouncing and walking droplets can behave just like quantum particles, even though the physics involved is entirely classical, and the mathematical models that reproduce and explain the behaviour are based on Newtonian mechanics alone. In 2006 Couder and Emmanuel Fort showed that the droplets mimic the double-slit experiment that so puzzled the founders of quantum theory.[91] They equipped the vibrating bath with analogues of two slits, and repeatedly walked a droplet towards them. The droplet either went through one slit or the other, just like a particle, and emerged in a direction that had a certain amount of random variation. But when they measured the positions of the emerging drops and plotted them as statistical histograms, the result was just like a diffraction pattern.

This casts doubt on Feynman's claim that the double-slit experiment has no classical explanation. Although Bush observed which slit the particle went through, using light, that's not the right analogue of a quantum observation, which must involve something much more energetic. In the spirit of Feynman's thought experiment, where he reckoned it's obvious that detecting the photon as it travels through a slit would mess up the diffraction patterns, we can be confident that an appropriate analogue of a quantum measurement – giving the droplet a serious wallop – would also mess up the diffraction pattern.

Other experiments provide further striking parallels with quantum mechanics. A droplet can hit a barrier that ought to stop it in its tracks, yet miraculously appear on the other side – a kind of quantum tunnelling, in which a particle can pass through a barrier even though it lacks the energy needed to do so. A pair of droplets can orbit round each other, much as an electron orbits a proton in a hydrogen atom.

However, unlike planets orbiting the Sun, the distances between the droplets are quantised: they occur as a series of specific discrete values, just like the usual quantisation of energy levels in an atomic nucleus, where only specific discrete energies are possible. A droplet can even follow a circular orbit on its own, giving it angular momentum – a rough analogue of quantum spin.

No one is suggesting that this particular classical fluid system explains quantum uncertainty. An electron probably isn't a microscopic droplet bouncing on a bath of cosmic fluid that fills space, and droplets don't match quantum particles in every detail. But this is just the simplest fluid system of its kind. It suggests that quantum effects seem weird because we're comparing them to the wrong classical models. If we think a particle is a tiny solid ball, and nothing else, or that a wave is like ripples on water, but nothing else, then wave/particle duality is indeed weird. Either it's a wave, or it's a particle, right?

The droplets tell us unequivocally that this is wrong. Maybe it could be both, interacting, and which aspect we seem to see depends on which features we're observing. We tend to focus on the droplet, but it's intimately bound up with its wave. There's a sense in which the droplet (call it 'particle' now for emphasis) tells the system how a particle ought to behave, but the wave tells it how a wave ought to behave. In the double-slit experiment, for instance, the *particle* goes though just one slit, but the *wave* goes through both. It's no surprise that in statistical averages over many trials, the pattern of waves shows up in the pattern of particles.

Could quantum mechanics really be like that, deep beneath the cold equations of the 'shut up and calculate' brigade?

It could be. But that's not a new theory.

MAX BORN DEVELOPED THE CURRENT interpretation of the quantum wave function of a particle in 1926. It doesn't tell us where the particle is located; it tells us the probability of it being in any given place. The only caveat in today's physics would be that a particle doesn't actually have a location; the wave function tells us the probability that an *observation* will locate it in a given position. Whether it was 'really' in that location before it was observed is at best philosophical speculation, and at worst a misunderstanding.

A year later, de Broglie suggested a reinterpretation of Born's idea, in which maybe the particle does have a location, but it can mimic a wave in suitable experiments. Perhaps the particle has an invisible companion, a 'pilot' wave that tells it how to behave like a wave. Essentially, he suggested that the wave function is a real physical object, and its behaviour is determined by Schrödinger's equation. The particle has a definite position at any time, so it follows a deterministic path, but it's *guided* by its wave function. If you have a system of particles, their combined wave function satisfies an appropriate version of Schrödinger's equation. The positions and momenta of the particles are hidden variables, which in conjunction with the wave function affect the outcomes of observations. In particular, the probability density of the positions is deduced from the wave function by following Born's prescription.

Wolfgang Pauli objected, saying that the pilot-wave theory didn't agree with certain phenomena in particle scattering. De Broglie couldn't give a satisfactory answer, off the top of his head, so he abandoned the idea. In any case, physicists could observe the probability distribution, but it seemed impossible to observe both the particle and its pilot wave at the same time. Indeed, it became conventional wisdom that the wave function isn't observable in its entirety – only bits of it, so to speak. When von Neumann offered his erroneous proof that no hidden-variable theory is possible, pilot waves sank without trace.

In 1952 a maverick physicist named David Bohm rediscovered the pilot-wave theory, and showed that Pauli's objection was unfounded. He developed a systematic interpretation of quantum theory as a deterministic system of pilot waves governed by hidden variables. He showed that all the standard statistical features of quantum measurements are still valid, so the pilot-wave theory agrees with the Copenhagen interpretation. The pictures show predictions for the double-slit experiment using Bohm's theory, and recent observations using weak measurements of individual photons, which don't disturb their states.[92] The resemblance is striking. By fitting smooth curves, it's even possible to deduce the probability distribution of the photons, recreating the diffraction pattern predicted by a wave model.

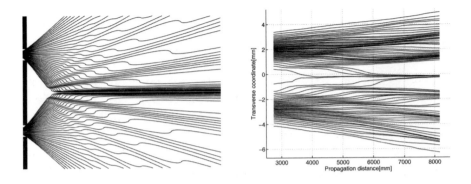

Left: Paths traced by electrons in the double-slit experiment according to Bohm's pilot-wave theory. *Right*: Experimental average paths using weak measurements of single photons.

Bohm's proposal didn't go down well with the quantum experts. One reason, unrelated to the physics, was that he'd been a communist in his youth. A better one was that the pilot-wave theory is by definition non-local. The behaviour of a system of particles, which are localised, depends on their combined wave function, which isn't. The wave function spreads throughout space, and it depends on the boundary conditions as well as the particles.

John Bell was more positive, and promoted the Bohm–de Broglie pilot-wave theory. Initially he wondered whether the non-locality could be removed. Eventually he came up with his famous proof that it can't be. Nevertheless, some physicists continued to work on non-local alternatives. This wasn't as stupid as it sounds. Let's see why.

ANY MATHEMATICAL THEOREM IS BASED on assumptions. It's an if/then statement: if certain assumptions are valid, then certain consequences must logically follow. The proof explains *how* they follow. The statement of the theorem ought to list all of the assumptions, but often some are tacit – those that are so standard in the field that there's no need for an explicit statement. Sometimes a close examination of the proof reveals that it hinges on some assumption that hasn't been stated and isn't standard. This represents a logical loophole, through which the theorem's consequences can escape.

Bell developed his theorem by spotting the same loophole in von Neumann's attempt that Hermann had found, and repairing it. But physicists can be very stubborn, and mathematicians can be very pedantic, so from time to time people look for unnoticed loopholes in Bell's theorem. Even if you find one, that doesn't of itself give a viable hidden-variable theory for quantum mechanics, but it's a hint that there might be one.

Tim Palmer, a physicist turned meteorologist who was still interested in physics, spotted one such loophole in 1995. Suppose there's a hidden dynamic which is deterministic but chaotic. He realised that if the dynamical system is sufficiently badly behaved, the proof of Bell's inequality falls apart, because the correlations it considers aren't computable. For example, suppose we want to model the spin of an electron. We've seen that the spin can be measured in any specified direction, and that (in suitable units) it's always 1/2 or −1/2. Which sign occurs seems to be random. Imagine that the hidden variables form a nonlinear dynamical system with two attractors, one corresponding to a spin of 1/2 and the other to a spin of −1/2. From given initial conditions, the spin evolves to one of those two values. Which? Each attractor has its own basin of attraction. If the variables start in one basin they're attracted to the spin 1/2 attractor; if they start in the other basin, they're attracted to the spin −1/2 attractor.

If the basins have fairly simple shapes, with nice boundaries, the proof of Bell's theorem works, and this two-attractor idea fails. However, basins of attraction can be very complicated. Chapter 10 mentioned that two (or more) attractors can have riddled basins, intertwined so intricately that the slightest disturbance switches states from one basin to the other one. Now the proof of Bell's theorem fails, because the correlations that it discusses don't exist as sensible mathematical objects.

'Uncomputable' here has a subtle meaning, and it doesn't stop nature using such a system. After all, the Copenhagen interpretation ('it just collapses, we don't know how') is even less computable, since it doesn't specify any mathematical process for the collapse. The statistical distribution of states ±1/2 should be related to statistical properties of the riddled basins, and those can be computable and meaningful, so comparison with experiment wouldn't be an issue.

Palmer supported this model with detailed calculations. He even suggested that the cause of the wave function's collapse might be gravity. Other physicists had previously entertained similar ideas, because gravity is nonlinear, destroying the superposition principle. In Palmer's model, gravity propels the state of the electron towards one or other attractor. Since then, Palmer has produced a series of papers examining other loopholes in Bell's theorem. They haven't yet led to a specific proposal for a hidden-variable dynamic that would base the whole of quantum mechanics on deterministic chaos, but they're valuable as explorations of what might work.

IN A SPECULATIVE SPIRIT, I want to suggest some other potential loopholes in Bell's theorem.

The proof of the theorem relies on comparing three correlations. By referring them to a presumed probability distribution on the space of hidden variables, Bell derives a relation between them, which happens to be an inequality. This relation is deduced by integrating the probability distribution over subsets of the space of hidden variables, determined by the correlations being measured. It can then be proved that these integrals are related in a manner that leads to the Bell inequality.

All very elegant – but what if the space of hidden variables doesn't have a probability distribution? Then the calculations used to prove the inequality have no meaning. A probability distribution is a special kind of measure, and many mathematical spaces don't have sensible measures. In particular, the space of all possible wave functions is usually infinite-dimensional – all combinations of infinitely many eigenstates. These are called Hilbert spaces, and they don't have sensible measures.

I should explain what I mean by 'sensible' here. Every space has at least one measure. Pick a point, which I'll call 'special'. Assign measure 1 to all subsets containing the special point, and 0 to the rest. This measure (said to be 'atomic', with the special point as the atom) has its uses, but it's not remotely like volume. To rule out such trivial measures, observe that the volume of an object in three-dimensional space doesn't change if you move the object sideways. This is called translation invariance. (It also doesn't change if you rotate the object, but I don't need that property.) The atomic measure just described isn't translation-invariant, because a translation can move the special point (measure 1) to some other location (measure 0). In a quantum context it would be natural to look for an analogous translation-invariant measure on Hilbert space. However, a theorem of George Mackey and André Weil tells us there's no such thing, except for rare cases when the Hilbert space happens to be finite-dimensional.

Now, although the entire space of hidden variables has no sensible probability measure, correlations of observations can still be meaningful. An observation is a projection from the space of wave functions to a single eigenstate, and each eigenstate lives in a finite-dimensional space, which does carry a measure. It therefore seems entirely reasonable that if a hidden dynamic for a quantum system exists, it should have an infinite-dimensional state space. After all, that's how the *non-hidden* variables work. In effect, the wave function *is* the hidden variable, 'hidden' because you can't observe it as a whole.

This isn't a new proposal. Lawrence Landau showed that the Einstein–Podolsky–Rosen experiment leads to Bell's inequality if you assume a hidden-variable theory based on a classical (Kolmogorov) probability space, but it doesn't if you assume an infinite number of independent hidden variables, because no such probability space exists. That's one loophole.

Another concerns the main objection to pilot waves, non-locality. The waves extend throughout the entire universe, and respond instantly to changes, however far away they may be. But I wonder if that objection is overstated. Think of the droplet experiments, which produce phenomena remarkably similar to quantum theory in a deterministic setting, and are an effective large-scale physical analogy for pilot waves. The droplet is arguably local. The corresponding wave isn't, but it certainly doesn't extend throughout the entire universe. It's confined to a dish. As far as explaining the double-slit experiment goes, all we need is a wave that extends *far enough* to notice the two slits. It can even be argued that without some sort of quasi-non-local 'halo', a photon can't know there's a slit to go through, let alone a choice of two. The model slit may be extremely thin, but the real one is a lot wider than a photon. (This is another mismatch between crisp boundary conditions and messy reality.)

A third loophole is the tacit assumption that the probability space of hidden variables is non-contextual, that is, independent of the observation being made. If the distribution of hidden variables depends on the observation, the proof of Bell's inequality fails. A non-contextual probability space might seem unreasonable: how can the hidden variables 'know' how they're going to be observed? If you toss a coin, the coin doesn't know it's going to hit a table until it does so. However, quantum observations get round Bell's inequality somehow, so the quantum formalism must allow correlations that violate the inequality. How? Because the quantum state is contextual. The measurements you make depend not only on the actual state of the quantum system – assuming it has one – but on the type of measurement. Otherwise the Bell inequality wouldn't be violated in experiments.

This isn't weird; it's entirely natural. I said that 'the coin doesn't know it's going to hit a table', but that's irrelevant. The coin doesn't *need* to know. It's not the state of the coin that provides the context, but the observation, which is an interaction between coin and table. The outcome depends on how the coin is observed, as well as the internal state of the coin. If for simplicity we imagine the coin spinning in zero gravity, and then we intercept it with a table, the result depends on when we do that, and on the angle between the spin axis of the coin and the plane of the table. Observed from a plane parallel to the axis, the coin alternates heads and tails. Observed from a plane perpendicular to the axis, it's spinning on its edge.

Since quantum wave functions are contextual, it seems only fair to allow hidden variables to be contextual too.

AS WELL AS PHILOSOPHICAL SPECULATION about the meaning of quantum phenomena, there's another reason for considering deterministic hidden-variable theories: the desire to unify quantum theory with relativity.[93] Einstein himself spent years searching – fruitlessly – for a unified field theory that combines quantum theory with gravity. Such a 'theory of everything' remains the holy grail of fundamental physics. The front-runner, string theory, has gone out of favour to some extent in recent years; the failure of the Large Hadron Collider to detect new subatomic particles predicted by superstrings probably hasn't done anything to reverse that. Other theories such as loop quantum gravity have their adherents, but nothing has yet emerged that satisfies mainstream physicists. Mathematically, there's a mismatch on a fairly basic level: quantum theory is linear (states can superpose) but general relativity isn't (they don't).

Most attempts to concoct a unified field theory leave quantum mechanics inviolate, and tinker with the theory of gravity to make it fit. In the 1960s, this approach very nearly worked. Einstein's basic equation for general relativity describes how the distribution of matter in a gravitational system interacts with the curvature of spacetime. There, the matter distribution is a crisp mathematical object with a crisp physical interpretation. In the semiclassical Einstein equation, the matter distribution is replaced by a quantum object defining the average matter distribution to be expected over many observations – a good guess at where the matter is, rather than a precise statement. This allows matter to be quantum while spacetime remains classical. As a working compromise, this variant of the Einstein equations had many successes, among them Stephen Hawking's discovery that black holes emit radiation. But it doesn't work so well when faced with the bugbear of how quantum observations behave. If the wave function suddenly collapses, the equations give inconsistent results.

Roger Penrose and Lajos Diósi independently tried to fix this problem in the 1980s, replacing relativity by Newtonian gravity. Anything learned in this version might with luck be extended to relativistic gravity. The problem with this approach, it emerged, is Schrödinger's cat, which shows up in an even more extreme way as Schrödinger's Moon. The Moon can split into two superposed pieces, one half circling the Earth and the other half somewhere else. Worse, the existence of such macroscopic superposed states allows signals to travel faster than light.

Penrose traced these failures to the insistence on not tinkering with quantum mechanics. Maybe it was to blame, not gravity. The key to the whole issue was the simple fact that even physicists who accept the Copenhagen interpretation can't actually tell us how the wave function collapses. It doesn't seem to happen if the measuring apparatus is itself a small quantum system, such as another particle. But if you measure the spin of a photon with standard apparatus, you get a specific result, not a superposition. How big does the measuring apparatus have to be to collapse the wave function of what it's observing? Why does sending a photon through a beam-splitter not disturb its quantum state, but sending it to a particle detector does? Standard quantum theory offers no answers.

This problem becomes acute when we're studying the entire universe. Thanks to the Big Bang theory of the origins of spacetime, the nature of quantum observations is a hugely important question for cosmology. If a quantum system's wave function collapses only when it's observed by something external, how can the wave function of the universe have collapsed to give all those planets, stars, and galaxies? That needs an observation from outside the universe. It was all rather confusing.

Some commentators on Schrödinger's cat have inferred, from the term 'observation', that observation requires an observer. The wave function collapses only when some conscious intelligent entity observes it. Therefore one reason humans exist may be that without us, the universe itself wouldn't exist, and this gives us a purpose in life and also has the merit of explaining why we're here. However, this train of thought gives humanity a privileged status, which seems arrogant and is one of the standard errors we keep making throughout the history of science. It also fits uncomfortably with the evidence that the universe has been around for about 13 billion years plus, apparently obeying the same general physical rules, without us being there to observe it. Moreover, this 'explanation' is bizarrely self-referential. Because we exist, we can observe the universe, causing it to exist ... which in turn causes us to exist. We're here because we're here because we're here. I'm not saying there are no ways to wriggle round such objections, but the whole idea gets the relationship between humanity and the universe back to front. We're here because the universe exists, not the other way round.

Less excitable personalities infer only that the wave function collapses when a small quantum system interacts with a sufficiently large one. Moreover, the large one behaves like a classical entity, so its wave function must already have collapsed. Maybe what's happening is just that sufficiently big systems collapse automatically.[94] Daniel Sudarsky is currently investigating one approach, spontaneous collapse. His view is that quantum systems collapse at random of their own accord, but when one particle collapses, it triggers collapse in all the others. The more particles, the more likely it is that one will collapse, and then they all do. So big systems become classical.

Maaneli Derakshani realised that a spontaneous collapse version of quantum theory might fit better with Newtonian gravity. In 2013 he discovered that the weird Schrödinger's Moon states go away when Newtonian gravity is combined with a spontaneous collapse theory. The first attempts still allowed signals to travel faster than light, however, which was bad. Part of the problem is that Newtonian physics doesn't automatically forbid such signals, unlike relativity. Antoine Tilloy is exploring a modified type of collapse which happens spontaneously in random locations in spacetime. As a consequence, previously fuzzy distributions of matter acquire specific locations, giving rise to gravity. Spacetime stays classical, so it can't do a Schrödinger's Moon. This gets rid of faster-than-light signals. The really big development would be to dump Newton and replace him by Einstein: to combine a collapse quantum theory with general relativity. Sudarsky's group is now attempting just that.

Oh, yes: I promised to explain about trampolining tardigrades. Gröblacher plans to test theories of quantum collapse by making a thin membrane spanning a square frame, a tiny trampoline a millimetre across. Make it vibrate, and use a laser to push it into a superposed state: part 'up' and part 'down'. Better still, sit a tardigrade on top, and see if you can superpose states of *that*.

Schrödinger's moss piglet. Love it!

IT GETS WEIRDER...

Most physicists believe that the formalism of quantum theory, Copenhagen and all, applies not just to electrons, moss piglets, and cats, but to any real-world system, however complex. But the latest take on Schrödinger's cat, published in 2018 by Daniela Frauchiger and Renato Renner[95], casts doubt on this belief. The difficulty shows up in a thought experiment in which physicists use quantum mechanics to model a system of physicists using quantum mechanics.

The basic idea goes back to 1967 when Eugene Wigner tweaked Schrödinger's scenario to argue that the orthodox quantum formalism can generate inconsistent descriptions of reality. He placed a physicist, 'Wigner's friend', in the box to observe the cat's wavefunction, revealing which of the two possible states the cat is in. An observer outside the box, however, still considers the cat to be in a superposition of 'alive' and 'dead', so the two physicists disagree about the cat's state. However, there's a flaw: Wigner's friend can't communicate his observation to the external observer, who can reasonably consider Wigner's friend to be in a superposition of 'observed dead cat' and 'observed live cat'. From the outside point of view, this state eventually collapses to just one of the two possibilities, but only when the box is opened. This scenario is different from what Wigner's friend has been thinking all along, but there's no logical inconsistency.

To get a genuine contradiction, Frauchiger and Renner go one step further. They use physicists instead of cats, which is obviously more ethical; it also permits a more complicated set-up. Physicist Alice sets the spin of a particle at random to either up or down, and sends the particle to her colleague Bob. He observes it when they and their labs are both inside boxes, and their states become entangled. Another physicist, Albert, models Alice and her lab using quantum mechanics, and the usual mathematics of entangled states implies that sometimes (not always!) he can deduce with complete certainty the state that Bob observed. Albert's colleague Belinda does the same for Bob and his lab, and the same mathematics implies that she can sometimes deduce with complete certainty the spin state that Alice set. Obviously this must be the state Bob measured. However, when you do the sums using the orthodox mathematics of quantum theory, it turns out that if this process is repeated many times, there must be a small percentage of cases in which Albert's and Belinda's deductions – both totally correct – disagree.

Ignoring the (rather complex) details, the paper focuses on three assumptions, all in accordance with orthodox quantum physics:

- The standard rules of quantum mechanics can be applied to any real-world system.
- Different physicists will not reach contradictory results by applying those rules correctly to the same system.
- If a physicist makes a measurement, its outcome is unique. For example, if she measures a spin to be 'up' then she can't also maintain (correctly) that it's 'down'.

Frauchiger and Renner's thought experiment proves a 'no-go' theorem:

it's impossible for all three statements to be true. Therefore the orthodox formalism of quantum mechanics is self-contradictory.

Quantum physicists have not received this news with great enthusiasm, and seem to be pinning their hopes on finding loopholes in the logic, but so far no one has found any. If the argument holds up, at least one of those three assumptions has to be abandoned. The most likely one to be sacrificed is the first, in which case physicists would have to accept that some real-world systems are beyond the scope of standard quantum mechanics. To deny the second or third assumption would be even more shocking.

As a mathematician, I can't help feeling that 'shut up and calculate' is in danger of missing something important. The reason is that if you do shut up, the calculations make sense. There are rules, often beautiful rules. They work. The mathematics behind them is deep and elegant, but it's built on a foundation of irreducible randomness.

So how does the quantum system know it should obey the rules?

I'm not alone in wondering whether there's a deeper theory that explains this, and I can't see any valid reason why it has to be irreducibly probabilistic. The droplet experiments aren't, for a start, but they sure do resemble quantum puzzles. The more we learn about nonlinear dynamics, the more it looks as though history would have been very different if we'd done that before discovering the quantum world.

David Mermin traces the 'shut up and calculate' mentality to World War II, where quantum physics was closely associated with the Manhattan project to develop atomic weapons. The military actively encouraged physicists to get on with the sums and stop worrying about their meaning. In 1976 Nobel-winning physicist Murray Gell-Mann said:[96] 'Niels Bohr brainwashed a whole generation of theorists into thinking that the job [of interpreting quantum theory] was done 50 years ago.' In *What is Real?*, Adam Becker suggests that the roots of this attitude lie in Bohr's insistence on the Copenhagen interpretation. As I've suggested, the insistence that only the results of experiments have meaning, and that there's no deeper underlying reality behind them, seems to have been an overreaction to logical positivism. Becker, like me, accepts that quantum theory *works*, but he adds that leaving it in its current state amounts to 'papering over a hole in our understanding of the world – and ignoring a larger story about science as a human process'.[97]

That doesn't tell us how to fill the hole, and neither does anything else in this chapter. Though there are hints. One thing is certain: if a deeper layer of reality *is* involved, we'll never find it by convincing ourselves it's not worth seeking.

17

EXPLOITING UNCERTAINTY

The best laid schemes o' mice an' men
Gang aft agley.
Robert Burns, *To a Mouse*

UNTIL NOW, I'VE GENERALLY DISCUSSED uncertainty as a problem; something that makes it difficult for us to understand what's likely to happen in the future, something that makes it possible for all of our best-laid plans to go 'agley', that is, wrong. We've investigated where the uncertainty comes from, what form it takes, how to measure it, and how to mitigate its effects. What I haven't done is to look at how we can use it. In fact, there are many circumstances in which a bit of uncertainty acts to our advantage. So although uncertainty is usually seen as a problem, it can also be a solution, though not always to the *same* problem.

THE MOST DIRECT USE OF RANDOMNESS occurs in the solution of mathematical problems that seem intractable by direct methods. Instead of simulating the solution and then sampling many runs to estimate the uncertainties involved, this approach turns the whole thing upside down, by making many sample simulations and inferring the solution from them. This is the Monte Carlo method, named after the famous casino.

The traditional toy example is finding the area of a complicated shape. The direct method is to cut the shape into pieces whose areas can be calculated by known formulas, and add the results together. More complex shapes can be handled using integral calculus, which essentially does the same thing, approximating the shape by lots of very thin rectangles. The Monte Carlo approach is very different. Put the shape inside a border whose area is known, say a rectangle. Throw lots of random darts at the shape, and count the proportion that hit it, compared to all the darts that hit the rectangle. If, say, the rectangle has area one square metre, and the darts hit the shape 72% of the time, its area must be somewhere around 0·72 square metres.

This method comes with a lot of small print. First, it's best when seeking a ballpark figure. The result is approximate, and we need to estimate the likely size of the error. Second, the darts need to be distributed uniformly over the rectangle. A good darts player, aiming at the shape, might hit it every time. What we want is a very poor darts player who sprays the darts all over the place, with no preferred direction. Third, bad estimates sometimes occur by chance. However, there are advantages too. The poor darts player can be arranged using tables of random numbers or, better still, computer calculations. The method works in higher dimensions – the volume of a complex three-dimensional space or a more conceptual 'volume' in many more dimensions. In mathematics, high-dimensional spaces abound, and they're not mysterious: just a geometric language for talking about large numbers of variables. Finally, it's usually far more efficient than a direct method.

Monte Carlo methods were invented (in the sense of being explicitly recognised as a general technique) by Stanislaw Ulam in 1946, while he was working on nuclear weapons at the Los Alamos National Laboratory in the USA. He was getting over an illness, passing the time by playing a form of patience called Canfield solitaire. Being a mathematician, he wondered whether he could work out the chance of winning by applying combinatorics and probability theory. Having tried and failed, he 'wondered whether a more practical method than "abstract thinking" might be to lay it out, say, one hundred times and simply observe and count the number of successful plays'.

The computers of the time were good enough to do the sums. But since he was also a mathematical physicist, Ulam immediately started wondering about the big problems that were holding back progress in nuclear physics, such as how neutrons diffuse. He realised that the same idea would provide practical solutions whenever a complicated differential equation could be reformulated as a random process. He passed the idea on to von Neumann, and they tried it on a real problem. It needed a code name, and Nicholas Metropolis suggested 'Monte Carlo', a favourite haunt of Ulam's gambling uncle.

Monte Carlo methods were vital to the development of the hydrogen bomb. From some points of view the world might well be a better place if Ulam had never had this insight, and I hesitate to advance nuclear weapons as a reason for doing mathematics. But it does illustrate the devastating power of mathematical ideas, and a powerful exploitation of randomness.

IRONICALLY, THE MAIN OBSTACLE TO the development of Monte Carlo methods was getting a computer to behave randomly.

Digital computers are deterministic. Give one a program, and it will carry out every instruction to the letter. This feature has led exasperated programmers to invent the spoof command DWIT (Do What I'm Thinking), and users to wonder about artificial stupidity. But this determinism also makes it hard for computers to behave randomly. There are three main solutions. You can engineer in some non-digital component that behaves unpredictably; you can provide inputs from some unpredictable real-world process such as radio noise; or you can set up instructions to generate pseudorandom numbers. These are sequences of numbers that appear to be random, despite being generated by a deterministic mathematical procedure. They're simple to implement and have the advantage that you can run exactly the same sequence again when you're debugging your program.

The general idea is to start by telling the computer a single number, the 'seed'. An algorithm then transforms the seed mathematically to get the next number in the sequence, and the process is repeated. If you know the seed and the transformation rule, you can reproduce the sequence. If you don't, it may be hard to find out what the procedure is. SatNav (the Global Positioning System) in cars makes essential use of pseudorandom numbers. GPS requires a series of satellites sending out timing signals, which are received by the gadget in your car, and analysed to work out where your car is. To avoid interference, the signals are sequences of pseudorandom numbers, and the gadget can recognise the right signals. By comparing how far along the sequence the message arriving from each satellite has got, it computes the relative time delays between all the signals. That gives the relative distances of the satellites, from which your position can be found using old-fashioned trigonometry.

In our post–chaos theory world, the existence of pseudorandom numbers is no longer paradoxical. Any chaotic algorithm can generate them. So can algorithms that technically aren't chaotic. Many practical ones eventually start repeating exactly the same sequence of numbers over and over again, but if it takes a billion steps before that happens, who cares? An early algorithm involved starting with a large integer seed, say

554,378,906

Now square it, getting

307,335,971,417,756,836

There are regular mathematical patterns in squares of numbers near both ends. For instance, here the final digit 6 happens again because $6^2 = 36$ also ends in 6. And you can predict that the first digit must be 3 because $55^2 = 3025$ starts with 3. Patterns of this kind aren't random, so to avoid them we chop off the ends (delete the rightmost six, then keep the next nine, say) to leave

335,971,417

Now square that to get

112,876,793,040,987,889

but keep only the middle chunk

876,793,040

Now repeat.

One theoretical issue with this recipe is that it's very hard to analyse it mathematically, to find out whether it really does behave like a random sequence. So different rules are generally used instead, the commonest being linear congruential generators. Here the procedure is to multiply the number by some fixed number, add another fixed number, but then reduce everything to its remainder on dividing by some specific big number. For efficiency, do everything in binary arithmetic. A big advance was the invention of the Mersenne twister in

1997 by Makoto Matsumoto and Takuji Nishimura. It's based on the number $2^{19,937} - 1$. This is a Mersenne prime – a prime number that is one less than a power of 2, a number-theoretic curiosity going back to the monk Marin Mersenne in 1644. In binary it consists of 19,937 consecutive 1s. The transformation rule is technical. The advantage is that the sequence of numbers it generates repeats only after $2^{19,937} - 1$ steps, which is a number with 6002 digits, and sub-sequences of up to 623 numbers are uniformly distributed.

Faster and better pseudorandom number generators have been developed since. The same techniques are useful for internet security, being used to encrypt messages. Each step in the algorithm can be viewed as 'encoding' the previous number, and the objective is to construct cryptographically secure pseudorandom number generators, which generate their numbers using a code that is provably difficult to break. The precise definition is more technical.

I'VE BEEN A BIT NAUGHTY.

Statisticians go to some lengths to explain that randomness is a feature of *processes*, not their outcomes. If you toss a fair dice ten times in a row, you *might* get 6666666666. In fact, on average that should happen once in every 60,466,176 trials.

However, there's a slightly different sense in which 'random' does sensibly apply to the outcomes, so that a sequence like 2144253615 is more random than 6666666666. The difference is best formulated for very long sequences, and it characterises a typical sequence created by a random process. Namely, all of the expected statistical features of the sequence should be present. Every digit 1–6 should appear roughly 1/6 of the time; every sequence of two successive digits should occur roughly 1/36 of the time, and so on. More subtly, there shouldn't be any long-range correlations; what happens in a given position and one two steps away, say, shouldn't repeat any pair of numbers significantly more often than any other. So we rule out something like 3412365452 because odd and even numbers alternate.

The mathematical logician Gregory Chaitin introduced the most extreme form of such conditions in his theory of algorithmic information. In conventional information theory, the amount of information in a message is the number of binary digits ('bits') needed to represent it. So a message 1111111111 contains ten bits of information, and so does 1010110111. Chaitin focused not on the sequence, but on the rules that can generate it – the algorithms to produce it. Those can also be coded in binary, in some programming language, precisely which isn't terribly important when the sequences get long enough. If the repeated 1s go on for, say, a million terms, then the sequence looks like 111...111 with a million 1s. The program 'write down 1 one million times', encoded into binary in any sensible way, is much shorter. The length of the shortest program that can generate a given output is the algorithmic information contained in that sequence. This description ignores some subtleties, but it will suffice here.

The sequence 1010110111 looks more random than 1111111111. Whether it is depends on how it continues. I chose it to be the first ten binary digits of π. Suppose it continues like that for a million digits. It will look extremely random. But the algorithm 'compute the first million binary digits of π' is much shorter than a million bits, so the algorithmic information in the million-digit sequence is a lot less than one million bits, even though π's digits satisfy all standard statistical tests for randomness. What they don't satisfy is 'different from the digits of π'. No one in their right mind would use the digits of π as an encryption system; the enemy would soon work it out. On the other hand, if the sequence were to continue in a genuinely random manner, instead of coming from π, it would probably be hard to find a shorter algorithm that generates it.

Chaitin defined a sequence of bits to be random if it's incompressible. That is, if you write down an algorithm to generate it up to a given position, then the algorithm is at least as long as the sequence, provided the number of digits gets very big. In conventional information theory, the amount of information in a binary string is the number of bits it contains – its length. The algorithmic information in a binary sequence is the length of the shortest algorithm that generates it. So the algorithmic information in a random sequence is also its length, but the algorithmic information in the digits of π is the length of the most compact program that generates it. That's much smaller.

Using Chaitin's definition, we can sensibly say that a specific sequence is random. He proved two interesting things about random sequences:

■ Random sequences of 0s and 1s exist. Indeed, almost every infinite sequence is random.

■ If a sequence is random, you can never prove it.

The proof of the first comes from counting how many sequences of given length there are, compared to shorter programs that might generate them. For instance, there are 1024 ten-bit sequences, but only 512 nine-bit programs, so at least half the sequences can't be compressed using shorter programs. The proof of the second is basically that if you can prove a sequence is random, the proof compresses the data in it, so it's not.

NOW, SUPPOSE YOU WANT TO generate a sequence, and you want to be convinced that it truly is random. Maybe you're setting up the key to some encryption scheme, for example. Apparently, Chaitin's work rules such a thing out. But in 2018 Peter Bierhorst and coworkers published a paper showing that you can get round this restriction using quantum mechanics.[98] Essentially, the idea is that quantum indeterminacy can be translated into specific sequences, with a physical guarantee that they're random in Chaitin's sense. That is, no potential enemy can deduce the mathematical algorithm that creates them – because there isn't one.

It might seem that the security of a random number generator can be guaranteed only if it satisfies two conditions. The user must know how the numbers are generated, otherwise they can't be sure that it generates truly random numbers. And the enemy must be unable to deduce the internal workings of the random number generator. However, there's no way to satisfy the first condition in practice using a conventional random number generator, because whatever algorithm it implements, it might go wrong. Keeping an eye on its internal workings might do the trick, but that's usually not practical. The second condition violates a basic principle of cryptography called Kerckhoff's principle: you must assume that the enemy knows how the encoding system works. Just in case they do. Walls have ears. (What you hope they don't know is the *decoding* system.)

Quantum mechanics leads to a remarkable idea. Assuming no deterministic hidden-variable theory exists, you can create a quantum-mechanical random number generator that is provably secure and random, such that the two conditions above both fail. Paradoxically, the user doesn't know anything about how the random number generator works, but the enemy knows this in complete detail.

The device uses entangled photons, a transmitter, and two receiving stations. Generate pairs of entangled photons with highly correlated polarisations Send one photon from each pair to one station, and the other to a second station. At each station, measure the polarisation. The stations are far enough apart that no signal can travel between them while they make this measurement, but by entanglement the polarisations they observe must be highly correlated.

Now comes the nifty footwork. Relativity implies that the photons can't be used as a faster-than-light communicator. This implies that the measurements, though highly correlated, must be unpredictable. The rare occasions on which they disagree must therefore be genuinely random. Violation of Bell inequalities, coming from entanglement, guarantees that the outcomes of these measurements are random. The enemy must agree with this assessment, whatever they know about the process used by the random number generator. The user can test for violations of Bell inequalities only by observing the statistics of the random number generator outputs; the internal workings are irrelevant for this purpose.

This general idea has been around for a while, but Bierhorst's team carried it out experimentally, using a set-up that avoids known loopholes in Bell inequalities. The method is delicate and the violations of Bell inequalities are slight, so it takes a long time to produce a sequence that's guaranteed to be random. Their experiment was like generating a random sequence of two equally likely outcomes by tossing a coin that generates heads 99.98% of the time. It can be done, by analysing the sequence after it has been generated. Here's one way. Run along the sequence until you reach the first point at which consecutive tosses are different: HT or TH. These pairs have the same probability, so you can consider HT to be 'heads' and TH to be 'tails'. If the probability of H is very big, or very small, you'll reject most of the data in the sequence, but what's left acts like a fair coin.

Running the experiment for ten minutes involved observing 55 million pairs of photons and led to a random sequence 1024 bits long. Conventional quantum random number generators, though not provably secure, generate millions of random bits per second. So right now, the extra security that this method guarantees isn't worth the hassle. Another problem is the size of the set-up: the two stations are 187 metres apart. This isn't a gadget you can carry in your briefcase, let alone put in a mobile phone. Miniaturising the set-up seems difficult, and putting it on a chip seems out of the question for the foreseeable future. Still, the experiment provides a proof of concept.

RANDOM NUMBERS (I'LL DROP THE 'pseudo' now) are used in a huge variety of applications. Innumerable problems in industry and related areas involve optimising some procedure to produce the best possible result. For example, an airline may wish to timetable its routes so that it uses the smallest number of aircraft, or to use a given number of aircraft to cover as many routes as possible. Or, more precisely, to maximise the profit that arises. A factory may wish to schedule maintenance of its machines to minimise the 'down time'. Doctors might want to administer a vaccine so that it can be most effective.

Mathematically, this kind of optimisation problem can be represented as locating the maximum value of some function. Geometrically, this is like finding the highest peak in a landscape. The landscape is usually multidimensional, but we can understand what's involved by thinking of a standard landscape, which is a two-dimensional surface in three-dimensional space. The optimum strategy corresponds to the position of the highest peak. How can we find it?

The simplest approach is hill climbing. Start somewhere, chosen however you wish. Find the steepest upward path and follow it. Eventually you'll reach a point where you can't go any higher. This is the peak. Well, maybe not. It's *a* peak, but it need not be the highest one. If you're in the Himalayas and climb the nearest mountain, it probably isn't Everest.

Hill climbing works well if there's only one peak, but if there are more, the climber can get trapped on the wrong one. It always finds a local maximum (nothing else nearby is higher), but maybe not a global one (*nothing* else is higher). One way to avoid getting trapped is to give the climber a kick every so often, teleporting them from one location to another one. If they're stuck on the wrong peak, this will get them climbing a different one, and they'll climb higher than before if the new peak is higher than the old one, *and* they don't get kicked off it too soon. This method is called simulated annealing, because of a metaphorical resemblance to the way atoms in liquid metal behave as the metal cools down, and finally freezes to a solid. Heat makes atoms move around randomly, and the higher the temperature, the more they move. So the basic idea is to use big kicks early on, and then reduce their size, as if the temperature is cooling down. When you don't initially know where the various peaks are, it works best if the kicks are random. So the right kind of randomness makes the method work better. Most of the cute mathematics goes into choosing an effective annealing schedule – the rule for how the size of kicks decreases.

ANOTHER RELATED TECHNIQUE, WHICH CAN solve many different kinds of problem, is to use genetic algorithms. These take inspiration from Darwinian evolution, implementing a simple caricature of the biological process. Alan Turing proposed the method in 1950 as a hypothetical learning machine. The evolutionary caricature goes like this. Organisms pass on their characteristics to their offspring, but with random variation (mutation). Those that are fitter to survive in their environment live to pass on their characteristics to the next generation, while the less fit ones don't (survival of the fittest or natural selection). Continue selecting for enough generations, and the organism gets very fit indeed – close to optimal.

Evolution can be modelled rather coarsely as an optimisation problem: a population of organisms wanders randomly around a fitness landscape, climbing the local peaks, and the ones that are too low down die out. Eventually the survivors cluster around a peak. Different peaks correspond to different species. It's far more complicated, but the caricature is sufficient motivation. Biologists make a big song and dance about evolution being inherently random. By this they mean (entirely sensibly) that evolution doesn't start out with a goal and aim towards it. It didn't decide millions of years ago to evolve human beings, and then keep choosing apes that got closer and closer to that ideal until it reached perfection in us. Evolution doesn't know ahead of time what the fitness landscape looks like. In fact, the landscape itself may change over time as other species evolve, so the landscape metaphor is somewhat strained. Evolution finds out what works better by testing different possibilities, close to the current one but randomly displaced. Then it keeps the better ones and continues the same procedure. So the organisms keep improving, step by tiny step. That way, evolution simultaneously constructs the peaks of the fitness landscape, finds out where they are, and populates them with organisms. Evolution is a stochastic hill-climbing algorithm, implemented in wetware.

A genetic algorithm mimics evolution. It starts with a population of algorithms that try to solve a problem, randomly varies them, and selects those that perform better than the rest. Do this again for the next generation of algorithms, and repeat until you're happy with the performance. It's even possible to combine algorithms in a parody of sexual reproduction, so that two good features, one from each of two different algorithms, can be united in one. This can be seen as a kind of learning process, in which the population of algorithms discovers the best solution by trial and error. Evolution can be seen as a comparable learning process, applied to organisms instead of algorithms.

There are hundreds of applications of genetic algorithms, and I'll mention just one to give a flavour of how they work. University timetables are highly complex. Hundreds of lectures must be scheduled so that thousands of students can follow innumerable different courses of study. It's common for students to take optional courses in different subjects; the American 'major' and 'minor' subject areas are a simple example. The lectures have to be scheduled to avoid clashes, where a student is required to listen to two lectures at the same time, and to avoid cramming three lectures on the same topic into consecutive slots. A genetic algorithm starts with a timetable and finds out how many clashes or triple-headers there are. Then it modifies the timetable randomly seeking a better one, and repeats. It might even be possible to combine parts of two fairly successful timetables together, mimicking sexual recombination.

SINCE WEATHER FORECASTING HAS INHERENT limitations, how about weather *control*? Don't ask whether rain is going to ruin the picnic or the invasion: make sure it doesn't.

Folk tales in Northern Europe maintained that discharging

cannons can prevent hailstorms. Anecdotal evidence for this effect appeared after several wars, including the Napoleonic war and the American Civil War: every time there's a big battle, it rains afterwards. (If that convinces you, I've got a really nice bridge, going cheap.) Towards the end of the 19th century the US Department of War spent $9000 on explosives and set them off in Texas. Nothing with any scientific validity was observed. Seeding clouds with silver iodide particles, which are very fine and in theory provide nuclei around which water vapour can condense, is widely used today to cause rain to fall, but when this works it's still not clear whether the rain would have fallen anyway. Several attempts have been made to weaken hurricanes by injecting silver iodide into their eyewalls, and again results have been inconclusive. The National Oceanographic and Atmospheric Administration of the USA has been exploring theoretical ideas to stop hurricanes, such as firing lasers at the storms that might give rise to them, triggering lightning discharges and dissipating some of the energy in the storms. And there are conspiracy theories that climate change isn't caused by human production of CO_2, but it's some sinister secret group controlling the weather to disadvantage America.

Randomness comes in many forms, and chaos theory tells us that a butterfly flap can radically change the weather. We've discussed the sense in which this is true: 'change' really means 'redistribute and modify'. When von Neumann was told of this effect, he pointed out that as well as making weather unpredictable, it potentially makes it controllable. To redistribute a hurricane, find the right butterfly.

We can't do this for a hurricane or a tornado. Not even a light drizzle. But we can do it for the electrical waves in a heart pacemaker, and it's widely used to plan fuel-efficient space missions when time isn't vital. In both cases, the main mathematical effort goes into selecting the right butterfly. That is, sorting out how, when, and where to interfere very slightly with the system to obtain the desired result. Edward Ott, Celso Grebogi, and James Yorke worked out the basic mathematics of chaotic control in 1990.[99] Chaotic attractors typically contain huge numbers of periodic trajectories, but these are all unstable: any slight deviation from one of them grows exponentially. Ott, Grebogi, and Yorke wondered whether controlling the dynamical system in the right way can stabilise such a trajectory. These embedded periodic trajectories are typically saddle points, so that some nearby states are initially attracted towards them, but others are repelled. Almost all of the nearby states eventually cease to be attracted and fall into the repelling regions, hence the instability. The Ott–Grebogi– Yorke method of chaotic control repeatedly changes the system by small amounts. These perturbations are chosen so that every time the trajectory starts to escape, it's recaptured: not by giving the state a push, but by modifying the system and moving the attractor, so the state finds itself back on the in-set of the periodic trajectory.

A human heart beats fairly regularly, a periodic state, but sometimes fibrillation sets in and the heartbeat becomes seriously irregular – enough to cause death if not stopped quickly. Fibrillation occurs when the regular periodic state of the heart breaks up to give a special type of chaos: spiral chaos, in which the usual series of circular waves travelling across the heart falls apart into a lot of localised spirals.[100] A standard treatment for irregular heartbeats is to fit a pacemaker, which sends electrical signals to the heart to keep its beats in sync. The electrical stimuli supplied by the pacemaker are quite large. In 1992 Alan Garfinkel, Mark Spano, William Ditto, and James Weiss reported experiments on tissue from a rabbit heart.[101] They used a chaotic control method to convert spiral chaos back into regular periodic behaviour, by altering the timing of the electrical pulses making the heart tissue beat. Their method restored regular beats using voltages far smaller than those in conventional pacemakers. In principle, a less disruptive pacemaker might be constructed along these lines, and some human tests were carried out in 1995.

Chaotic control is now common in space missions. The dynamical feature that makes it possible goes right back to Poincaré's discovery of chaos in three-body gravitation. In the application to space missions, the three bodies might be the Sun, a planet, and one of its moons. The first successful application, proposed by Edward Belbruno in 1985, involved the Sun, Earth, and the Moon. As the Earth circles the Sun and the Moon circles the Earth, the combined gravitational fields and centrifugal forces create an energy landscape with five stationary points where all the forces cancel out: one peak, one valley, and three saddles. These are called Lagrange points. One of them, L_1, sits between the Moon and the Earth, where their gravitational fields and the centrifugal force of the Earth circling the Sun cancel out. Near this point, the dynamics is chaotic, so the paths of small particles are highly sensitive to small perturbations.

A space probe counts as a small particle here. In 1985 the International Sun–Earth Explorer ISEE-3 had almost completely run out of the fuel used to change its trajectory. If it could be transported to L_1 without using up much fuel, it would be possible to exploit the butterfly effect to redirect it to some distant objective, still using hardly any fuel. This method allowed the satellite to rendezvous with comet Giacobini–Zinner. In 1990 Belbruno urged the Japanese space agency to use a similar technique on their probe, *Hiten*, which had used up most of its fuel completing its main mission. So they parked it in a lunar orbit and then redirected it to two other Lagrange points to observe trapped dust particles. This kind of chaotic control has been used so often in unmanned space missions that it's now a standard technique when fuel efficiency and lower costs are more important than speed.

18

UNKNOWN UNKNOWNS

There are known knowns; there are things we know we know. We also know there are known unknowns; that is to say we know there are some things we do not know. But there are also unknown unknowns – the ones we don't know we don't know.
Donald Rumsfeld, *US Department of Defense news briefing, 12 February 2002*

RUMSFELD MADE HIS FAMOUS REMARK to advocate invading Iraq even though there was no evidence linking it to al-Qaeda's terrorist attack on Manhattan's World Trade Center towers in 2001. His 'unknown unknowns' were what *else* Iraq might be up to, even though he had no idea what it might be, and that was being presented as a reason for military action.[102] But in a less militaristic context, the distinction he was making might well have been the most sensible thing he said in his entire career. When we're aware of our ignorance, we can try to improve our knowledge. When we're ignorant but unaware of it, we may be living in a fool's paradise.

Most of this book has been about how humanity took unknown unknowns and turned them into known unknowns. Instead of attributing natural disasters to the gods, we recorded them, pored over the measurements, and extracted useful patterns. We didn't obtain an infallible oracle, but we did get a statistical oracle that forecast the future better than random guesswork. In a few cases, we turned unknown unknowns into known knowns. We knew how the planets were going to move, and we also knew *why* we knew that. But when our new oracle of natural law proved inadequate, we used a mixture of experiment and clear thinking to quantify uncertainty: although we were still uncertain, we knew how uncertain we were. Thus was probability theory born.

My six Ages of Uncertainty run through the most important advances in understanding why we're uncertain, and what we can do about it. Many disparate human activities played a part. Gamblers joined forces with mathematicians to unearth the basic concepts of probability theory. One of the mathematicians *was* a gambler, and at first he used his mathematical knowledge to great effect, but eventually he lost the family fortune. The same story goes for the world's bankers early this century, who so firmly believed that mathematics freed their gambles from risk that they also lost the family fortune. Their family was a bit bigger, though: the entire population of the planet. Games of chance presented mathematicians with fascinating questions, and the toy examples these games provided were simple enough to analyse in detail. Ironically, it turns out that neither dice nor coins are as random as we imagine; much of the randomness comes from the human who throws the dice or tosses the coin.

As our mathematical understanding grew, we discovered how to apply the same insights to the natural world, and then to ourselves. Astronomers, trying to obtain accurate results from imperfect observations, developed the method of least squares to fit data to a model with the smallest error. Toy models of coin tossing explained how errors can average out to give smaller ones, leading to the normal distribution as a practical approximation to the binomial, and the central limit theorem, which proved that a normal distribution is to be expected when large numbers of small errors combine, whatever the probability distribution of the individual errors might be.

Meanwhile, Quetelet and his successors were adapting the astronomers' ideas to model human behaviour. Soon, the normal distribution became the statistical model *par excellence*. An entire new subject, statistics, emerged, making it possible not just to fit models to data, but to assess how good the fit was, and to quantify the significance of experiments and observations. Statistics could be applied to anything that could be measured numerically. The reliability and significance of the results were open to question, but statisticians found ways to estimate those features too. The philosophical issue 'what is probability?' caused a deep split between frequentists, who calculated probabilities from data, and Bayesians, who considered them to be degrees of belief. Not that Bayes himself would necessarily sign up to the viewpoint now named after him, but I think he'd be willing to take credit for recognising the importance of conditional probability, and for giving us a tool to calculate it. Toy examples show what a slippery notion conditional probability can be, and how poorly human intuition can perform in such circumstances. Its real-world applications, in medicine and the law, sometimes reinforce that concern.

Effective statistical methods, used on data from well-designed clinical trials, have greatly increased doctors' understanding of diseases, and made new drugs and treatments possible by providing reliable assessments of their safety. Those methods go well beyond classical statistics, and some are feasible only because we now have fast computers that can handle huge amounts of data. The financial world continues to pose problems for all forecasting methods, but we're learning not to place too much reliance on classical economics and normal distributions. New ideas from areas as disparate as complex systems and ecology are shedding new light and suggesting sensible policies to ward off the next financial crash. Psychologists and neuroscientists are starting to think that our brains run along Bayesian tracks, embodying beliefs as strengths of connections between nerve cells. We've also come to realise that sometimes uncertainty is our friend. It can be exploited to perform useful tasks, often very important ones. It has applications to space missions and heart pacemakers.

It's also why we're able to breathe. The physics of gases turned out to be the macroscopic consequences of microscopic mechanics. The statistics of molecules explains why the atmosphere doesn't all pile up in one place. Thermodynamics arose from the quest for more efficient steam engines, leading to a new and somewhat elusive concept: entropy. This in turn appeared to explain the arrow of time, because entropy increases as time passes. However, the entropy explanation on a macroscopic scale conflicts with a basic principle on the microscopic scale: mechanical systems are time-reversible. The paradox remains puzzling; I've argued that it comes from a focus on simple initial conditions, which destroy time-reversal symmetry.

Round about the time when we decided that uncertainty isn't the whim of the gods, but a sign of human ignorance, new discoveries at the frontiers of physics shattered that explanation. Physicists became convinced that in the quantum world, nature is irreducibly random, and often plain weird. Light is both a particle and a wave. Entangled particles somehow communicate by 'spooky action at a distance'. Bell inequalities are guarantees that only probabilistic theories can explain the quantum world.

About sixty years ago, mathematicians threw a spanner in the works, discovering that 'random' and 'unpredictable' aren't the same. Chaos shows that deterministic laws can produce unpredictable behaviour. There can be a prediction horizon beyond which forecasts cease to be accurate. Weather-forecasting methods have changed completely as a result, running an ensemble of forecasts to deduce the most probable one. To complicate everything, some aspects of chaotic systems may be predictable on much longer timescales. Weather (a trajectory on an attractor) is unpredictable after a few days; climate (the attractor itself) is predictable over decades. A proper understanding of global warming and the associated climate change rests on understanding the difference.

Nonlinear dynamics, of which chaos is a part, is now casting doubt on some aspects of Bell inequalities, which are now under assault through a number of logical loopholes. Phenomena once thought to be characteristic of quantum particles are turning up in good old classical Newtonian physics. Maybe quantum uncertainty isn't uncertain at all. Maybe Chaos preceded Cosmos as the ancient Greeks thought. Maybe Einstein's dictum that God doesn't play dice needs to be revised: He does play dice, but they're hidden away, and they're not truly random. Just like real dice.

I'm fascinated by how the Ages of Uncertainty are still striking sparks off each other. You often find methods from different ages being used in conjunction, such as probability, chaos, and quantum, all rolled into one. One outcome of our long quest to predict the unpredictable is that we now *know* there are unknown unknowns. Nassim Nicholas Taleb wrote a book about them, *The Black Swan*, calling them black swan events. The 2nd-century Roman poet Juvenal wrote (in Latin) 'a rare bird in the lands and very much like a black swan', his metaphor for 'non-existent'. Every European *knew* swans were always white, right up to the moment in 1697 when Dutch explorers found black ones, in quantity, in Australia. Only then did it become apparent that what Juvenal thought he knew wasn't a known known at all. The same error dogged the bankers in the 2008 crisis: their 'five-sigma' potential disasters, too rare even to consider, turned out to be commonplace, but in circumstances they hadn't previously encountered.

All six Ages of Uncertainty have had lasting effects on the human condition, and they're still with us today. If there's a drought, some of us pray for rain. Some try to understand what caused it. Some try to stop everyone making the same mistake again. Some look for new sources of water. Some wonder if we could create rain to order. And some seek better methods to predict droughts by computer, using quantum effects in electronic circuits.

Unknown unknowns still trip us up (witness the belated realisation that plastic rubbish is choking the oceans), but we're beginning to recognise that the world is much more complex than we like to imagine, and everything is interconnected. Every day brings new discoveries about uncertainty, in its many different forms and meanings. The future is uncertain, but the science of uncertainty is the science of the future.

NOTES

1 The quote probably has nothing to do with Berra, and may come from an old Danish proverb:
https://quoteinvestigator.com/2013/10/20/no-predict/

2 *Ezekiel* 21:21.

3 Ray Hyman. Cold reading: how to convince strangers that you know all about them, *Zetetic* 1 (1976/77) 18–37.

4 I'm not sure exactly what's meant by 'lighted carbon' – maybe charcoal? – but several sources mention it in this connection, including:
John G. Robertson, *Robertson's Words for a Modern Age* (reprint edition), Senior Scribe Publications, Eugene, Oregon, 1991.
http://www.occultopedia.com/c/cephalomancy.htm

5 If 'beat the odds' means 'improve your chance of getting the winning numbers', then probability theory predicts that no such system works except by accident. If it means 'maximise your winnings if you *do* win', you can take some simple precautions. The main one is: avoid choosing numbers that a lot of other people will also be likely to choose. If your numbers come up (just as likely as any other set of numbers) you'll share the winnings with fewer people.

6 A good example is the wave equation, originally deduced from a model of a violin string as a line segment vibrating in the plane. This model paved the way for more realistic ones, used today for everything from analysing the vibrations of a Stradivarius to calculating the internal structure of the Earth from seismic recordings.

7 I'm aware that the singular is technically 'die', but nowadays pretty much everyone uses 'dice' in the singular as well. The word 'die' is old-fashioned and easily misinterpreted; many people are unaware that 'dice' is its plural. So in this book I'll write 'a dice'.

8   We ditched fate to make dice fairer, *New Scientist*, 27 January 2018, page 14.

9   If you're happy about the red/blue dice, but not convinced this is right when the dice look identical, two things may help. First, how do the coloured dice 'know' to produce twice as many combinations as they would have done if they had the same colour? That is, how can the colours affect the throws to that extent? Second: take two dice so similar that even you can't tell which is which, throw them a lot of times, and count the proportion of times you get two 4s. If unordered pairs decide the result, you'll get something close to 1/21. If it's ordered pairs, it should be close to 1/36.
    If you're not convinced even about coloured dice, the same considerations apply, but you should do the experiment with coloured dice.

10   The 27 ways to total 10 are:

$$1+3+6 \quad 1+4+5 \quad 1+5+4 \quad 1+6+3$$
$$2+2+6 \quad 2+3+5 \quad 2+4+4 \quad 2+5+3 \quad 2+6+2$$
$$3+1+6 \quad 3+2+5 \quad 3+3+4 \quad 3+4+3 \quad 3+5+2 \quad 3+6+1$$
$$4+1+5 \quad 4+2+4 \quad 4+3+3 \quad 4+4+2 \quad 4+5+1$$
$$5+1+4 \quad 5+2+3 \quad 5+3+2 \quad 5+4+1$$
$$6+1+3 \quad 6+2+2 \quad 6+3+1$$

The 25 ways to total 9 are:

$$1+2+6 \quad 1+3+5 \quad 1+4+4 \quad 1+5+3 \quad 1+6+2$$
$$2+1+6 \quad 2+2+5 \quad 2+3+4 \quad 2+4+3 \quad 2+5+2 \quad 2+6+1$$
$$3+1+5 \quad 3+2+4 \quad 3+3+3 \quad 3+4+2 \quad 3+5+1$$
$$4+1+4 \quad 4+2+3 \quad 4+3+2 \quad 4+4+1$$
$$5+1+3 \quad 5+2+2 \quad 5+3+1$$
$$6+1+2 \quad 6+2+1$$

11   https://www.york.ac.uk/depts/maths/histstat/pascal.pdf

12   The ratio in which the stakes should be divided is

$$\sum_{k=0}^{s-1}\binom{r+s-1}{k} \quad \text{to} \quad \sum_{k=s}^{r+s-1}\binom{r+s-1}{k}$$

where player 1 needs r more rounds to win, player 2 needs s more. In this case, the ratio is:

$$\binom{8}{0} + \binom{8}{1} + \binom{8}{2} + \binom{8}{3} + \binom{8}{4} + \binom{8}{5} \quad \text{to}$$
$$\binom{8}{6} + \binom{8}{7} + \binom{8}{8}.$$

13 Persi Diaconis, Susan Holmes, and Richard Montgomery. Dynamical bias in the coin toss, *SIAM Review* **49** (2007) 211–235.

14 M. Kapitaniak, J. Strzalko, J. Grabski, and T. Kapitaniak. The three-dimensional dynamics of the die throw, *Chaos* **22** (2012) 047504.

15 Stephen M. Stigler, *The History of Statistics*, Harvard University Press, Cambridge, Massachusetts, 1986, page 28.

16 We want to minimise $(x - 2)^2 + (x - 3)^2 + (x - 7)^2$. This is quadratic in x and the coefficient of x^2 is 3, which is positive, so the expression has a unique minimum. This occurs when the derivative is zero, that is, $2(x - 2) + 2(x - 3) + 2(x - 7) = 0$. So $x = (2 + 3 + 7)/3$, the mean. A similar calculation leads to the mean for any finite set of data.

17 The formula is $\sqrt{2/n\pi}\exp[-2(x - \frac{1}{2}n)^2/n]$ considered as an approximation to the *probability* of getting x heads in n tosses.

18 Think of people entering the room one at a time. After k people have entered, the probability that all their birthdays are *different* is

$$\frac{365}{365} \times \frac{364}{365} \times \frac{363}{365} \times \cdots \times \frac{365 - k + 1}{365}$$

because each new arrival has to avoid the previous $k - 1$ birthdays. This is 1 minus the probability of at least one common birthday, so we want the smallest k for which this expression is *less than* 1/2. This turns out to be $k = 23$. More details are at:

https://en.wikipedia.org/wiki/Birthday_problem

19 Non-uniform distributions are discussed in:
M. Klamkin and D. Newman. Extensions of the birthday surprise, *Journal of Combinatorial Theory* **3** (1967) 279–282.

A proof that the probability of two matching birthdays is least for a uniform distribution is given in:
D. Bloom. A birthday problem, *American Mathematical Monthly* 80 (1973) 1141–1142.

20 The diagram looks similar but now each quadrant is divided into a 365×365 grid. The dark strips in each quadrant contain 365 squares each. But there is an overlap of 1 inside the target region. So this contains $365 + 365 - 1 = 729$ dark squares, there are $365 + 365 = 730$ dark squares outside the target, so the total number of dark squares is $729 + 730 = 1459$. The conditional probability of hitting the target is 729/1459, which is 0·4996.

21 For calculations, Quetelet used a binomial distribution for 1000 coin tosses, which he found more convenient, but he emphasised the normal distribution in his theoretical work.

22 Stephen Stigler, *The History of Statistics*, Harvard University Press, Cambridge, Massachusetts, 1986, page 171.

23 This isn't necessarily true. It assumes all distributions are obtained by combining bell curves. But it was good enough for Galton's purposes.

24 The word 'regression' came from Galton's work on heredity. He used the normal distribution to explain why, on the whole, children with either two tall parents or two short parents tend to be somewhere in between, calling this 'regression to the mean'.

25 Another figure deserving mention is Francis Ysidro Edgeworth. He lacked Galton's vision but was a far better technician, and put Galton's ideas on a sound mathematical basis. However, his story is too technical to include.

26 In symbols:

$$P\left(\left(X_1 + \cdots + \frac{X_n}{n} - \mu\right) < \beta\sqrt{n}\right) \rightarrow \int_{-\infty}^{\beta} e^{-y^2/2} dy$$

where the right-hand side is the cumulative normal distribution for mean 0 and variance 1.

27 We have $P(A|B) = P(A\&B)/P(B)$ and also $P(B|A) = P(B\&A)/P(A)$. But the event $A\&B$ is the same as $B\&A$. Dividing one equation by

the other, we have $P(A|B)/P(B|A) = P(A)/P(B)$. Now multiply both sides by $P(B|A)$.

28 Frank Drake introduced his equation in 1961 to summarise some of the key factors that affect the likelihood of alien life, as part of the first meeting of SETI (Search for ExtraTerrestrial Intelligence). It's often used to estimate the number of alien civilisations in the Galaxy, but many of the variables are difficult to estimate and it's not suitable for that purpose. It also involves some unimaginative modelling assumptions. See:
https://en.wikipedia.org/wiki/Drake_equation

29 N. Fenton and M. Neil. *Risk Assessment and Decision Analysis with Bayesian Networks*, CRC Press, Boca Raton, Florida, 2012.

30 N. Fenton and M. Neil. Bayes and the law, *Annual Review of Statistics and Its Application* 3 (2016) 51–77.

https://en.wikipedia.org/wiki/Lucia_de_Berk

31 Ronald Meester, Michiel van Lambalgen, Marieke Collins, and Richard Gil. On the (ab)use of statistics in the legal case against the nurse Lucia de B. arXiv:math/0607340 [math.ST] (2005).

32 The science historian Clifford Truesdell is reputed to have said: 'Every physicist knows what the first and the second law [of thermodynamics] mean, but the problem is that no two agree about them.' See:
Karl Popper. Against the philosophy of meaning, in: *German 20th Century Philosophical Writings* (ed. W. Schirmacher), Continuum, New York, 2003, page 208.

33 You can find the rest at:
https://lyricsplayground.com/alpha/songs/f/firstandsecondlaw.html.

34 N. Simanyi and D. Szasz. Hard ball systems are completely hyperbolic, *Annals of Mathematics* 149 (1999) 35–96.
N. Simanyi. Proof of the ergodic hypothesis for typical hard ball systems, *Annales Henri Poincaré* 5 (2004) 203–233.
N. Simanyi. Conditional proof of the Boltzmann–Sinai ergodic hypothesis. *Inventiones Mathematicae* 177 (2009) 381–413.
There is also a 2010 preprint, which seems not to have been published:
N. Simanyi. The Boltzmann–Sinai ergodic hypothesis in full

generality:
https://arxiv.org/abs/1007.1206

35 Carlo Rovelli. *The Order of Time*, Penguin, London 2018.

36 The figure is a computer calculation, also subject to the same errors. Warwick Tucker found a computer-aided but rigorous proof that the Lorenz system has a chaotic attractor. The complexity is real, not some numerical artefact.
W. Tucker. The Lorenz attractor exists. *C.R. Acad. Sci. Paris* **328** (1999) 1197–1202.

37 Technically, the existence of invariant measures that give the right probabilities has been proved only for special classes of attractors. Tucker proved the Lorenz attractor has one, in the same paper. But extensive numerical evidence suggests they're common.

38 J. Kennedy and J.A. Yorke. Basins of Wada, *Physica* D **51** (1991) 213–225.

39 P. Lynch. *The Emergence of Numerical Weather Prediction*, Cambridge University Press, Cambridge, 2006.

40 Fish later said that the caller was referring to a hurricane in Florida.

41 T.N. Palmer, A. Döring, and G. Seregin. The real butterfly effect, *Nonlinearity* **27** (2014) R123–R141.

42 E.N. Lorenz. The predictability of a flow which possesses many scales of motion. *Tellus* **3** (1969) 290–307.

43 T.N. Palmer. A nonlinear dynamic perspective on climate prediction. *Journal of Climate* **12** (1999) 575–591.

44 D. Crommelin. Nonlinear dynamics of atmospheric regime transitions, PhD Thesis, University of Utrecht, 2003.
D. Crommelin. Homoclinic dynamics: a scenario for atmospheric ultralow-frequency variability, *Journal of the Atmospheric Sciences* **59** (2002) 1533–1549.

45 The sums go like this:

total over 90 days	$90 \times 16 = 1440$
total over 10 days	$10 \times 30 = 300$
total over all 100 days	1740
average	$1740/100 = 17{\cdot}4$

which is 1·4 larger than 16.

46 For the 800,000-year record:
    E.J. Brook and C. Buizert. Antarctic and global climate history viewed from ice cores, *Nature* **558** (2018) 200–208.

47 This quotation appeared in *Reader's Digest* in July 1977, with no documentation. The *New York Times* published an article 'How a "difficult" composer gets that way' by the composer Roger Sessions on 8 January 1950. It included: 'I also remember a remark of Albert Einstein, which certainly applies to music. He said, in effect, that everything should be as simple as it can be, but not simpler!'

48 Data from the U.S. Geological Survey show that the world's volcanoes produce about 200 million tons of CO_2 per year. Human transportation and industry emit 24 billion tons, 120 times as big. https://www.scientificamerican.com/article/earthtalks-volcanoes-or-humans/

49 The IMBIE team (Andrew Shepherd, Erik Ivins, and 78 others). Mass balance of the Antarctic Ice Sheet from 1992 to 2017, *Nature* **558** (2018) 219–222.

50 S.R. Rintoul and 8 others. Choosing the future of Antarctica, *Nature* **558** (2018) 233–241.

51 J. Schwartz. Underwater, *Scientific American* (August 2018) 44–55.

52 E.S. Yudkowsky. An intuitive explanation of Bayes' theorem: http://yudkowsky.net/rational/bayes/

53 W. Casscells, A. Schoenberger, and T. Grayboys. Interpretation by physicians of clinical laboratory results, *New England Journal of Medicine* **299** (1978) 999–1001.
    D.M. Eddy. Probabilistic reasoning in clinical medicine: Problems and opportunities, in: (D. Kahneman, P. Slovic, and A. Tversky, eds.), *Judgement Under Uncertainty: Heuristics and Biases*, Cambridge University Press, Cambridge, 1982.
    G. Gigerenzer and U. Hoffrage. How to improve Bayesian reasoning without instruction: frequency formats, *Psychological Review* **102** (1995) 684–704.

54 The Kaplan–Meier estimator deserves mention, but it would interrupt the story. It's the most widely used method for estimating survival rates from data in which some subjects may leave the trial

before the full time period – either through death or other causes. It's non-parametric and second on the list of highly cited mathematics papers. See:
E.L. Kaplan and P. Meier. Nonparametric estimation from incomplete observations, *Journal of the American Statistical Association* **53** (1958) 457–481.
https://en.wikipedia.org/wiki/Kaplan%E2%80%93Meier_estimator

55 B. Efron. Bootstrap methods: another look at the jackknife, *Annals of Statistics* **7** B (1979) 1–26.

56 Alexander Viktorin, Stephen Z. Levine, Margret Altemus, Abraham Reichenberg, and Sven Sandin. Paternal use of antidepressants and offspring outcomes in Sweden: Nationwide prospective cohort study, *British Medical Journal* **316** (2018); doi: 10.1136/bmj.k2233.

57 Confidence intervals are confusing and widely misunderstood. Technically, the 95% confidence interval has the property that the true value of the statistic lies inside that interval for 95% of the times a confidence interval is calculated from a sample. It does *not* mean 'the probability that the true statistic lies in the interval is 95%.'

58 A corporate euphemism for 'these people will never be able to pay us back'.

59 Technically it was the Swedish National Bank's Prize in Economic Sciences in Memory of Alfred Nobel, set up in 1968, not one of the prize categories Nobel established in his 1895 will.

60 Technically, a distribution $f(x)$ has fat tails if it decays like a power law; that is, $f(x) \sim x^{-(1+a)}$ as x tends to infinity, for $a > 0$.

61 Warren Buffett. Letter to the shareholders of Berkshire Hathaway, 2008:
http://www.berkshirehathaway.com/letters/2008ltr.pdf

62 A.G. Haldane and R.M. May. Systemic risk in banking ecosystems, *Nature* **469** (2011) 351–355.

63 W.A. Brock, C.H. Hommes, and F.O.O. Wagner. More hedging instruments may destabilise markets, *Journal of Economic Dynamics and Control* **33** (2008) 1912–1928.

64 P. Gai and S. Kapadia. Liquidity hoarding, network externalities,

and interbank market collapse, *Proceedings of the Royal Society* A (2010) **466**, 2401–2423.

65 For a long time it was thought that the human brain contains ten times as many glial cells as neurons. Credible internet sources still say about four times. But a 2016 review of the topic concludes that there are slightly fewer glial cells than neurons in the human brain. Christopher S. von Bartheld, Jami Bahney, and Suzana Herculano-Houze, The search for true numbers of neurons and glial cells in the human brain: A review of 150 years of cell counting, *Journal of Comparative Neurology, Research in Systems Neuroscience* **524** (2016) 3865–3895.

66 D. Benson. Life in the game of Go, *Information Sciences* **10** (1976) 17–29.

67 Elwyn Berlekamp and David Wolfe. *Mathematical Go Endgames: Nightmares for Professional Go Players*, Ishi Press, New York 2012.

68 David Silver and 19 others. Mastering the game of Go with deep neural networks and tree search, *Nature* **529** (1016) 484–489.

69 L.A. Necker. Observations on some remarkable optical phaenomena seen in Switzerland; and on an optical phaenomenon which occurs on viewing a figure of a crystal or geometrical solid, *London and Edinburgh Philosophical Magazine and Journal of Science* **1** (1832) 329–337.
J. Jastrow. The mind's eye, *Popular Science Monthly* **54** (1899) 299–312.

70 I. Kovács, T.V. Papathomas, M. Yang, and A. Fehér. When the brain changes its mind: Interocular grouping during binocular rivalry. *Proceedings of the National Academy of Sciences of the USA* **93** (1996) 15508–15511.

71 C. Diekman and M. Golubitsky. Network symmetry and binocular rivalry experiments, *Journal of Mathematical Neuroscience* **4** (2014) 12; doi: 10.1186/2190-8567-4-12.

72 Richard Feynman, in a lecture: 'The Character of Physical Law'. Earlier Niels Bohr said 'Anyone who is not shocked by quantum theory has not understood it,' but that's not quite the same message.

73 Richard P. Feynman, Robert B. Leighton, and Matthew Sands. *The

Feynman Lectures on Physics, Volume 3, Addison-Wesley, New York, 1965, pages 1.1–1.8.

74 Roger Penrose. Uncertainty in quantum mechanics: Faith or fantasy? *Philosophical Transactions of the Royal Society* A **369** (2011) 4864–4890.

75 https://en.wikipedia.org/wiki/Complex_number

76 François Hénault. Quantum physics and the beam splitter mystery: https://arxiv.org/ftp/arxiv/papers/1509/1509.00393.

77 If the spin quantum number is n, the spin angular momentum is $S = (h/4\pi)\sqrt{n(n+2)}$, where h is Planck's constant.

78 Electron spin is curious. A superposition of two spin states $\uparrow$ and $\downarrow$ that point in opposite directions can be interpreted as a single spin state with an axis whose direction is related to the proportions in which the original states are superposed. However, a measurement about *any* axis yields either 1/2 or −1/2. This is explained in Penrose's paper cited in Note 74.

79 An unexamined assumption here is that if a classical cause produces a classical effect, then a quantum fraction of that cause (in some superposed state) produces a quantum fraction of the same effect. A half-decayed atom creates a half-dead cat. It makes some sort of sense in terms of probabilities, but if it were true in general, a half-photon wave in a Mach–Zehnder interferometer would create half a beam-splitter when it hits one. So this kind of superposition of classical narratives can't be how the quantum world behaves.

80 I discussed Schrödinger's cat at length in *Calculating the Cosmos*, Profile, London, 2017.

81 Tim Folger. Crossing the quantum divide, *Scientific American* **319** (July 2018) 30–35.

82 Jacqueline Erhart, Stephan Sponar, Georg Sulyok, Gerald Badurek, Masanao Ozawa, and Yuji Hasegawa. Experimental demonstration of a universally valid error-disturbance uncertainty relation in spin measurements, *Nature Physics* **8** (2012) 185–189.

83 Lee A. Rozema, Ardavan Darabi, Dylan H. Mahler, Alex Hayat, Yasaman Soudagar, and Aephraim M. Steinberg. Violation of Heisenberg's measurement-disturbance relationship by weak

measurements, *Physics Review Letters* **109** (2012) 100404. Erratum: *Physics Review Letters* **109** (2012) 189902.

84 If you generate a pair of particles, each having nonzero spin, in such a way that the total spin is zero, then the principle of conservation of angular momentum (another term for spin) implies that their spins will remain perfectly anticorrelated if they are then separated – as long as they're not disturbed. That is, their spins always point in opposite directions along the same axis. If you now measure one, and collapse its wave function, it acquires a definite spin in a definite direction. So the other one must also collapse, and give the opposite result. It sounds mad, but it seems to work. It's also a variation on my pair of spies; they've just antisynchronised their watches.

85 See Note 74.

86 Even male insects feel pleasure when they 'orgasm', *New Scientist*, 28 April 2018, page 20.

87 J.S. Bell. On the Einstein Podolsky Rosen paradox, *Physics* **1** (1964) 195–200.

88 Jeffrey Bub has argued that Bell and Hermann misconstrued von Neumann's proof, and that it does not aim to prove that hidden variables are completely impossible.
Jeffrey Bub. Von Neumann's 'no hidden variables' proof: A re-appraisal, *Foundations of Physics* **40** (2010) 1333–1340.

89 Adam Becker, *What is Real?*, Basic Books, New York 2018.

90 Strictly speaking, Bell's original version also requires the outcomes on both sides of the experiment to be exactly anticorrelated whenever the detectors are parallel.

91 E. Fort and Y. Couder. Single-particle diffraction and interference at a macroscopic scale, *Physical Review Letters* **97** (2006) 154101.

92 Sacha Kocsis, Boris Braverman, Sylvain Ravets, Martin J. Stevens, Richard P. Mirin, L. Krister Shalm, and Aephraim M. Steinberg. Observing the average trajectories of single photons in a two-slit interferometer, *Science* **332** (2011) 1170–1173.

93 This section is based on Anil Anathaswamy, Perfect disharmony, *New Scientist*, 14 April 2018, pages 35–37.

94 It can't just be size, can it? Consider the effect of a beam-splitter (a

1/4 phase shift) and a particle detector (scramble the wave function). Both are comfortably macroscopic. The first thinks it's quantum, the second knows it's not.

95 D. Frauchiger and R. Renner. Quantum theory cannot consistently describe the use of itself, *Nature Communications* (2018) 9:3711; doi: 10.1038/S41467-018-05739-8.

96 A. Sudbery. *Quantum Mechanics and the Particles of Nature*, Cambridge University Press, Cambridge, 1986, page 178.

97 Adam Becker, *What is Real?*, Basic Books, New York, 2018.

98 Peter Bierhorst and 11 others. Experimentally generated randomness certified by the impossibility of superluminal signals, *Nature* **223** (2018) 223–226.

99 E. Ott, C. Grebogi, and J.A. Yorke. Controlling chaos, *Physics Review Letters* **64** (1990) 1196.

100 The occurrence of chaos in heart failure, rather than just randomness, has been detected in humans:

Guo-Qiang Wu and 7 others, Chaotic signatures of heart rate variability and its power spectrum in health, aging and heart failure, *PLos Online* (2009) 4(2): e4323; doi: 10.1371/journal.pone.0004323.

101 A. Garfinkel, M.L. Spano, W.L. Ditto, and J.N. Weiss. Controlling cardiac chaos, *Science* **257** (1992) 1230–1235.
A more recent article on chaos control of a model heart is in:
B.B. Ferreira, A.S. de Paula, and M.A. Savi. Chaos control applied to heart rhythm dynamics, *Chaos, Solitons and Fractals* **44** (2011) 587–599.

102 At the time, President George W. Bush decided not to attack Iraq in response to 9/11. But soon after, the USA and its allies did invade, citing Saddam Hussein's 'support for terrorism' as a reason. The *Guardian* newspaper for 7 September 2003 reported a poll showing that 'seven out of ten Americans continue to believe that Saddam Hussein had a role' in the attacks, despite there being no evidence for this.
https://www.theguardian.com/world/2003/sep/07/usa.theobserver

PICTURE CREDITS

Page 11: David Aikman, Philip Barrett, Sujit Kapadia, Mervyn King, James Proudman, Tim Taylor, Iain de Weymarn, and Tony Yates. Uncertainty in macroeconomic policy-making: art or science? *Bank of England paper*, March 2010.

Page 141: Tim Palmer and Julia Slingo. Uncertainty in weather and climate prediction, *Philosophical Transactions of the Royal Society* A **369** (2011) 4751–4767.

Page 198: I. Kovács, T.V. Papathomas, M. Yang, and A. Fehér. When the brain changes its mind: Interocular grouping during binocular rivalry, *Proceedings of the National Academy of Sciences of the USA* **93** (1996) 15508–15511.

Page 237 (left figure): Sacha Kocsis, Boris Braverman, Sylvain Ravets, Martin J. Stevens, Richard P. Mirin, L. Krister Shalm, and Aephraim M. Steinberg. Observing the average trajectories of single photons in a two-slit interferometer, *Science* (3 Jun 2011) **332** issue 6034, 1170–1173.

INDEX

Note: An *italic* page reference indicates that relevant information appears only in an illustration. Where text on the same subject is already indexed, accompanying illustrations are not distinguished.